D0323424

IN THIS WE ARE NATIVE

Other Books by Annick Smith:

Headwaters: Montana Writers on Water and Wilderness

Big Bluestem: Journey into the Tallgrass

Homestead

The Last Best Place: A Montana Anthology (co-editor)

IN THIS WE ARE NATIVE

MEMOIRS AND JOURNEYS

ANNICK SMITH

The Lyons Press

Guilford, Connecticut
An Imprint of The Globe Pequot Press

Printed in the United States of America

Designed by a Good Thing, Inc.

10 9 8 7 6 5 4 3 2 1

Library of Congress Cataloging-in-Publication Data

The following essays have appeared in different forms in magazines, journals, and anthologies:
Audubon: Anticipating Loss; Sink or Swim; The Whales Are Sinking; Thanksgiving
The New York Times: Summering Ground; Pablo Neruda's House in the Sand; Titicaca, Rock of the Mountain Cat
Outside: Palo Duro Canyon
The *Missoula Independent:* Three Ways of Walking the Blackfoot
Islands: Margitsiziget, Where It All Began
Writing Down the River: Falling into the Canyon
Shadow Cat: Wildcats I Have Not Known
Headwaters: Sink or Swim
The Sweet Breathing of Plants and *Orion* magazine: Gleaning Wild Gardens
Big Sky Journal: The Best Last Place
Big Bluestem: Journey into the Tallgrass: Big Bluestem

For Jessyca Rose

Alex, Andrew, Stephen, Eric
Helene, Becca, Robin

and Bill

CONTENTS

PART 1

ANTICIPATING LOSS

The meadow is a cupped hand that holds me. It was cut from rock by a great glacial flood, and the clay of its bed undulates like waves. My meadow is clothed with native fescues, bunchgrasses, sedges, wild roses, and was seeded with European timothy, brome, and bluegrass. The land tilts down the Bear Creek watershed toward the Big Blackfoot River and is hemmed in by hills and cliffs of ponderosa, jack pine, and Douglas fir, splashed yellow in autumn with deciduous larch. Prevailing westerlies blow over the Garnet Range and funnel into the grass. After squalls, rainbows arc over the east ridge, framing far-off wilderness peaks.

Our recycled two-story hewn-log house rests at four thousand feet in a hay field across a dirt track from the log bunkhouse we call the little house. Both are chinked in white, zebra striped like my favorite Italian cathedrals—a hint of Tuscany in darkest Montana. Behind the bunkhouse and above the creek's canyon sits an oversized cement-block garage, and next to it a red shack we once used as a study. Overlooking the lower meadow is a log barn more than a century old with a new red tin roof, and farther downhill is a mouse-infested icehouse that was my boys' clubhouse. Called the chalet because of its geometry, it is slowly slipping into the root cellar below.

Ours is the last place on Bear Creek Road, out of sight and above the cabins, doublewides, and ranchettes of neighbors. I see history out every window. In the 1880s our 163 acres were homesteaded by Swedish immigrants who chopped down wide-girthed yellow pines in the natural meadows, used crosscut saws to level the stumps they could not pull out with workhorses, dug a gravity-flow irrigation system fed by a flume from Bear Creek, and pulled rocks from the clay soil, stacking them in piles that rise like ancient cairns.

After the Swedes came a clan of Yugoslavs, who cross-fenced the meadows, planted grains, and added rocks to the stone piles. In 1948 a ranching family from eastern Montana raised cattle here, and hay. They logged and did what they could to feed eight kids on what had by then become a starvation outfit. When we arrived in 1971, eager for land, they sold with no regrets.

My husband, Dave, made elaborate plans to restore the meadows and flume, build ponds for migrating waterfowl, plant orchards. He died before his plans could be made real, and under my laissez-faire occupation the meadow has gone to seed. My twin sons, Alex and Andrew, tease me by calling the place Entropy Ranch, but that's all right by me. The old homestead has become a haven for grazing cows, white-tailed deer, nesting snipe, and bluebirds; for serviceberry and thorn apple brush, blue camas, white daisies, anthills, and knapweed; for ski tracks in heavy snow, children playing ball, running dogs, motley horses—and for me, watching years pass like geese on the fly.

For most of my adult life I have found peace in Montana, far from the industrialized urban landscapes that had been my European and midwestern heritage. This has been my paradise. Often, I wake to the moonsongs of coyotes. Walking paths in the deep woods, I might encounter elk or the hulking, breath-stopping forms of bears. But paradise is too close to perfection for humans to endure, so we are compelled to populate our Edens with serpents both actual and virtual.

I stand on the porch watching a house being built at the meadow's edge, where no house has ever stood, and my illusion of wildness is broken. For the first time in thirty years, I will see the dwelling of a neighbor, and I am mad as a dehived bee. I wish I could dream up words powerful enough to fight suburban growth on one side, intense logging on the other, and some seventy miles to the east, near the headwaters of the Blackfoot River, the dangers of huge gold and copper mining projects that threaten the whole valley.

"Progress. You can't stop progress," says Evelyn, my downstream neighbor. I tell her with more bravado than truth, "Oh yes I can."

The first thing I've done is put a conservation easement on our ranch through the Montana Land Reliance. I live at the end of a three-mile gravel track off Highway 200. My quarter section is the only piece of individually owned ground in this part of the township that has not been broken up or platted for subdivision, and I am content that it never will be.

My four sons and their families will be able to live on this land if they choose, or sell it in one piece. Aside from the historic log outbuildings, only three livable dwellings may exist within the compound, but if one of the family wants privacy, the easement allows an additional house to be built in the woods at the southeastern corner of the property.

If this arrangement causes family squabbles, I won't care because I'll be dead. I plan to die here and have most of my ashes scattered over the meadow (reserve a handful, please, to be sprinkled in the Blackfoot River), but as long as I can think and speak, my business will be to protect the land. The easement states that our timber may never be cut except to enhance the rebirth of an old-growth forest. We will preserve the meadow and hay fields, allowing farming or grazing for domestic animals, but will not permit commercial enterprises such as a dude ranch, hunting camp, or golf course. Whoever will live here may conduct a cottage enterprise, as I do, but can do nothing to diminish or destroy the environmental qualities of the place. There will be no mining, drilling, or industrial manufacture.

I have far less influence over what happens outside my boundaries. The abrupt edge of cliffs above my southern fence line is state land on which I hold a grazing lease. Here, tall larches and yellow pines have remained unharvested for the thirty years my family has inhabited the place. A few years ago a state forester was preparing an environmental impact statement for nearby timber sales. We toured the range with an eye toward wildlife habitat, and he told me the state might cooperate in preserving that leased plot as part of a wildlife corridor that could run from the mouth of Bear Creek up to its spring-fed sources in the foothills.

Now I hesitate to approach state foresters. "Wise Use" Republicans in control of Montana's government have reduced standards for water quality and hacked away at environmental safeguards written into the Montana Constitution. Those same legislators recently ordered the Department of State Lands to accelerate logging and mining to provide income for public schools and universities. The last thing I want is to direct their avaricious eyes at my leased patch of mature timber.

Our cliffs remain unharmed, but tomorrow I may hear the whine of chain saws, the crash of big trees falling. Which is why, when I walk on a path made golden with needles from a roof of shedding larch, or stop to meditate in the dim understory of ponderosa pine, I am overwhelmed with a sense of transience. I hear the drumming of a grouse, the bugle of an elk in rut, the *tap-tap-tap* of a pileated woodpecker. I hold my breath. I feel grief coming on.

Anticipation of loss is devastating, but I keep climbing up the ridge, trying to focus on what is here and alive. I scramble onto a lichen-encrusted promontory. Below lies the smooth cup of our meadow. In all directions a checkered range extends like a crazy quilt over hills and divides, spreading into the Rattlesnake and Swan Wilderness Areas, threading along the blue-green Blackfoot River, stretched across the Garnet Range to the Clark Fork drainage. Over the years I have witnessed the blue-black density of these woods become whiskery as a badly razored beard. I have seen history at work in an "industrial" forest.

These thousands of evergreen hillsides were given in the 1800s in alternate sections to railway companies as a prize for opening up the West, then sold to the Anaconda Company, which logged them to build the shafts and tunnels underpinning their deep copper mines in Butte. Anaconda's worst practice was clear-cutting. Before the outcry against the technique began during the 1970s, they cut patches down to bare ground, starting in valleys where access was easy, building roads as they moved up the ridges, but leaving some of the interior forests unroaded and therefore wild. Since then many clear-cuts have come back strong and green, but they are returning as dense stands of one age and one or two species, increasing fire danger and diminishing biological diversity.

When Anaconda sold out to Champion Timberlands, Champion continued clear-cutting, but in the face of public outrage soon concentrated on what it called selective logging—selecting most of the marketable trees, a practice called high-grading. As I walked the logging roads with my dogs, seasoned loggers and log-truck drivers would complain to me about overcutting. "It's bad," they would say. "They're gonna take everything and get out. Won't be nothing left in a couple of years."

Champion sold the huge range (860,000 acres) to the Plum Creek Timber Company in 1992. It was like selling a corpse to vultures. Plum Creek has a history in these parts of being the most environmentally careless of all the logging outfits—the Darth Vader of the woods, according to an article in the *Wall Street Journal*. In a recent report to the Securities Exchange Commission the company said it expected to complete its logging of all mature and old-growth timber by 2000. They're doing a bang-up job, cutting small-girthed trees as well as big ones, on schedule and on target.

I'm a de facto expert on the work of these timber companies because their logging rigs and skidders, road graders and cats have rolled past my house for thirty years. Unfortunately, they have a right-of-way, a logging easement sold by the previous owners to Anaconda, then transferred to Champion and now Plum Creek. Logging roads wind from my gates north along the Bear Creek drainage to the Blackfoot River, west up Dirty Ike, and through the Game and Norman Creek watersheds into the Garnets.

When Plum Creek took over, company officials rode in on their silver 4x4 pickups proclaiming a new agenda. A pricey national advertising campaign declared they would abide by a set of principles that "protects and enhances the environmental values of the forests." Objectives included working with state and federal agencies "to protect and enhance all wildlife species"; protecting "water quality and fisheries by providing effective buffers along streams, other bodies of water and wetlands"; and cooperating with "neighboring landowners in addressing and minimizing potential cumulative effects."

I was as cynical as an Indian responding to promises from the government. So I was not surprised when Plum Creek accelerated logging in the steep, deep-cover elk range above my place and farther on around Bonner and Olson Peaks to proportions beyond anything I had seen before. From 1992 till now a nearly steady run of trucks loaded with high-country timber has passed my house bound for mills in Bonner and Libby and markets as far away as Japan. It's hard to believe there are any stands left uncut, but still they keep rolling. I am often awakened by trucks in the 3 A.M. dark, and disturbed by them at suppertime as the drivers stop to open my gate.

Bad news, but worse was to come. My neighbors and I learned that Plum Creek's next move would be to log Bear Creek canyon right through the canopied, park-like wildlife corridor that was our last most beautiful patch of old growth. Contract loggers would cut larch, fir, and ponderosa ranging from one hundred to three hundred years in age. They would log along my fence line and down the steep hillsides to the creek bottoms. This was not virgin forest, for large stumps and remains of an old sawmill tell tales of long-ago logging, but these woods still rose in splendor compared to the ravaged hillsides above them.

The company's timber managers assured us that the logging would be done in environmentally sound ways, and it was—more or less—because we watched their every move.They built only one small new road. Hauled some trees up the draws with cables from line machines; skidded others out with tractors in winter. But environmentally sound was not good enough. Plum Creek took more than 50 percent of the volume of mature timber in the canyon and along the creek, transforming the watershed for at least the two hundred years it will take for old growth to return.

The one good thing that came from our fight to save Bear Creek's old growth was a uniting of watershed dwellers. Strange how a stream can tie folks together. Neighbors who barely know each other joined in a protest meeting with an ice-eyed manager of Plum Creek's forests.We never believed we could stop the whole timber sale, but we hoped to convince the company to leave the creek bottoms alone.

My neighbors are not environmental activists.Allen is a retired driver for Eddy's Bread, and his wife, Evelyn, delivered mail on rural routes. Their daughter works for the post office, and her then-husband was a drug and alcohol counselor at a backcountry "boot camp" for juvenile offenders. Other neighbors include a cabinetmaker, his wife, and baby; a fifty-some-year-old grandmother who is a folksinger and makes New Age arts and crafts and her younger husband, who constructs doors at a Missoula shop; the head of the carpenter's union; a teacher and a hospice nurse; and an elderly widow who summers in an A-frame near Bear Creek's mouth. We have never gotten together in a political way except for asking Plum Creek—please—to reconsider its plans.

Our watershed coalition talked its way up the corporate ladder and walked the managers along the meandering creek through tall trees striped for cutting in Day-Glo orange. My neighbors and I are not against logging. We know that foresters at the bottom of the ladder can't leave a good stand uncut for fear of losing their jobs. We'd have been more sympathetic if some of them, whom we've dealt with for years, had admitted reluctance. The bosses, as expected, never budged from their planned harvest.

What the company wants—what all corporations must strive for—is as much immediate profit as possible. What we want is sustainable timber management that will enable our sawyer friends to work in the woods until they're too creaky to heft a saw. We think the timber around here could support the next generation of loggers, and generations to follow, while remaining a healthy habitat for the flora and fauna that also depend on these woods.

"We need to open up the canopy," the company's senior forester told me, "give the young trees more sun to grow upward."

"No you don't," I said. "You could leave this one small watershed alone."

Opening the canopy was exactly what we feared. The canopy created that silent, roofed, life-enhancing forest that I, at least, loved enough to call sacred. A few summers ago a philosopher friend of the twins married his beloved in a wildflower clearing in the deep woods beside the creek. A hundred guests walked the half mile from my house to join in the ceremony. Children played in the rilling waters while the rest of us found refuge from the August heat in the green shade of high, moss-hung firs.

This is what we have lost: The tall, moss-hung firs are mostly cut down. The clearing lies under a mess of crackling dead limbs and treetops. In autumn dozens of multihued species of mushrooms used to push out of the damp earth of the clearing in the dim understory of creek-watered giant trees. Now there is knapweed in the sun. June brought rare fairy slipper orchids to shaded north-facing slopes, but the shade is gone and the Russian thistles are coming in. Just before the logging began, I spotted two black bears in the mossy, fern-licked bottoms. Now there is no cover for them, and all I see is their scat.

It is true that whitetails and elk still descend in spring (the logging has created more browse), but the mothers must go elsewhere to bed their young. A couple of great horned owls continue to nest in the remaining tall trees and snags, but their habitat is diminished. This is also true for the red-tailed hawks, and for the blue herons that nested in a rookery bordering Holbrook's creek-fed pond. I don't know what effect the logging has had on mountain lions and bobcats, whose tracks in the winter bottoms told us they were near. The wildcats prey on grouse, wild turkeys, snowshoe rabbits, squirrels, chipmunks, and other rodents that, like the prolific coyotes, need cover but don't seem to care if it's old growth.

Before loggers cut down its canopy, Bear Creek's bottoms were nearly free of invading Eurasian forbs such as knapweed, leafy spurge, and sulphur cinquefoil, which are a scourge all over Montana. That's because the canyon was dark and damp and mossy, no good for the nonnative weeds to grow, and because logging had not occurred there for more than a quarter century.

Log trucks carry the invasive seeds in their tire treads, and the seeds plant themselves in soil disturbed by skidders and other logging equipment. Once they take hold, it's good-bye native flora. Yellow-blooming leafy spurge, almost immortal with roots twenty feet deep, has crept into the Blackfoot valley from Missoula, and knapweed with its lavender flower is poisoning the region, devouring grasses that animals and birds depend on.

The situation causes our state wildlife manager for this area to shake his head, even though he is employed by an administration friendly to natural resource extraction.

"Used to be," he told me, "you'd fly over the Garnets and see logging on a different watershed each year. Now we're seeing Plum Creek logging every watershed at the same time."

Logged ground, contrary to popular opinion, may be more prone to fire than pristine areas. When lightning strikes, logged areas burn because they are filled with dry brush and slash, dense regrowth and weak trees. During the summer of 2000 forest fires burned a huge hunk of western Montana in a season hotter and drier than any on record. Ashes fell from white skies. When winds blew north from the flaming Bitterroots, even untouched valleys like ours were choked under a shawl of smoke.

All of us who live near woods were scared. We watered and mowed and packed valuables in pickups ready for a quick getaway. If fire came, we were warned, there'd be no fighters to save us, for they were occupied elsewhere. We knew our skimpy second- and third-growth forests would be lost. And with most of the fire-hardened, genetically superior big trees cut down, regeneration would likely take a long time. In my case, beyond a lifetime.

This was true all over the West, and can be seen clearly from the air. Green-black swaths of wildlife-sheltering tall timber are rare. Where once were forests, you see a few wooded blotches on ridges and canyons, and next to them vibrant green squares of dense new growth on clear-cuts, or tarred ground with black skeleton trees. Under your wings lie pale, road-laced hills, mountains dotted with feathery seed trees and slash. The scarred, scorched, weedy earth spreads for hundreds of miles.

My battle with Plum Creek has recently become exceedingly personal. The company has claimed about four acres inside the fence on the southeastern corner of our land. They surveyed the property line, staked it, hung it with glow-in-the-dark pink ribbons. It doesn't matter that for thirty years we believed this land was ours. That we have cherished and preserved it, naming the patch Bear Cove because it was a haven for black bears. It will be a haven no more.

A young woman who is the company's field forester in this region tells me there are not enough harvestable trees on those few acres to make the expense of surveying, refencing, and logging worthwhile. But the old hands who are her bosses want the property line neat and clear, she says. They won't consider selling me the land. "It's not negotiable," she says. "I tried."

Bad blood with this gal is not an issue. She's a recent graduate of the University of Montana's forestry school, educated in more environmentally sound methods of timber management, and I like her. "How could you take this job," I ask, "if you care about preserving forests?"

"People like me are your best hope for the future, she says.

I pray she's right. Meanwhile I'm so upset I forget to mention being a good neighbor. Anaconda and Champion had the manners and political sense to allow buffers to exist between their logging operations and neighboring properties. Why not Plum Creek?

This corporation, it seems, doesn't care about common people too poor to offer profitable deals. They cover their tracks with media smokescreens, promises, and politicians, being spinmeisters with an eye on national policies. On the ground, however, Plum Creek has reconfigured itself into a real-estate operation that is currently selling off prime parcels to developers, land trusts, public agencies, or any outfit that's willing to pay top dollar for logged ground on lakes or streams or mountain-view slopes. At the same time, the company is shifting its timber business to warmer, better tree-growing climates in Arkansas and South America, and buying more productive forests in Maine.

All of which fuels my paranoia, giving me reasons to believe that their incursion onto my turf is a kind of vendetta. Plum Creek doesn't really need my few acres of timber, but I've been a wasp and a gadfly—a pain in their corporate ass. By claiming my land and cutting my trees, they know they will break my heart.

Lucky for my sanity, the news is not all bad. Efforts to enhance the Blackfoot's fishery are a more successful story. A coalition called the Blackfoot Challenge—uniting environmental groups with the state's Department of Fish, Wildlife, and Parks as well as federal agencies, companies such as Plum Creek, and locals—has been working to preserve endangered bull trout and threatened native cutthroats in the river and its tributaries.

Bear Creek is a spring-fed tributary that was once prime breeding grounds for native fish, but fish numbers are down. My neighbors and I have been cooperating with the state's fisheries experts and, yes, Plum Creek, to restore pools and spawning grounds. The creek needed help because a section above its mouth had been moved from its bed to the edge of the hills to expand pasturage and real estate. "What you had," said the fisheries expert "was a 1,021-yard riffle."

Young trout struggle in that much fast water, so the state brought in heavy machinery to relocate the stream to its historic bed. They dug holes in the riffle, creating pools. Then workers dropped logs into the pools to provide cover for fish and to channel water into the banks, where new meanders will form. They planted willows to provide shade and cool water, which trout need to survive. Now there is some variation in stream levels, and places where fine gravels can accrete. Soon, we hope, Bear Creek will again hold spawning beds for bull trout and cutts.

Plum Creek helped restore the creek because it is part of the company's bull trout recovery plan. By taking the initiative, it hopes to preempt the EPA from imposing stricter regulations in watersheds such as ours. It also replaced culverts under logging roads with wider culverts that are supposed to allow fish to pass. And, sad for my downstream neighbors, they diverted the creek to run around the dammed pond near the heron rookery (a fishing spot for both humans and herons), because the lunker brown trout that hide there devour young bull trout and native cutts. The pond may eventually become the wetland marsh it was before bulldozers dug it out, a healthier environment for the fishery and for the rare tailed frogs that live in its mud.

With frogs disappearing all over the world, it is reassuring that Bear Creek harbors this survivor. Tailed frogs are an indicator of good water quality. Gray or dark brown, one to two inches long, they often live in cool, swift high-elevation streams. They are vulnerable to sediment, for during their three-year metamorphosis the tadpoles must cling with specially adapted mouths to rocks in rapidly flowing waters. Their presence in our creek is a sign of its health.

Another part of Plum Creek's bull trout recovery

strategy has been to install locked gates at the entrances to its roads. They say this is to prevent buildup of silt from road traffic. We, who have used those roads for access to hunting, wood gathering, trail riding, mushrooming, berrying, and just plain hiking, can only scoff at their hypocrisy.

"Sure, lock the barn after the cows've escaped," I say. If the corporation wants to stop erosion, it should stop stripping the forest, cutting new roads, running heavy equipment and log trucks down steep slopes, leaving the ripped-open topsoil to run off with melting snow and rain. That's where the silt is coming from. Not from a handful of wood cutters and hunters.

My neighbors and I think the gates have been put up to lock out snoops like us, so the company can log the backcountry unobserved. But what's coming in and going out passes before our eyes whether we choose to look or not. Our best hope is to keep the damage down—which is what we try to do, knowing the effort is about as effective as trying to change the wind.

There are places in these woods I will not walk (and there are more every season) because the anticipation of loss has turned into loss itself. When we came to this valley, the wild country seemed enormous, if not endless. Now it seems to be passing away. In moods of self-pity I feel constricted, in jail, but then I look with a stranger's eye at what remains and ask myself how a person of conscience can complain about this much space and freedom?

Usually I don't complain. Over thirty years the roots

I've put down in Blackfoot clay are as deep as the roots of the scrawny weeping willow in my yard, which grows where no willow should grow, but hangs on. Realtors come by every so often to see if I'm ready to sell, but I wouldn't trade my worked-over ground and the scarred hills around it for anything. Not for wild mountains and scenic river valleys for sale all over the northern Rockies as vacation homes and exclusive fly-in clubs. Not for a million bucks.

When I walk in the woods these days, I look for new growth and try to fight off the despair that rides shotgun to lack of power. The forest is resilient. If people take enough care, it will come back like the bald eagles that are nesting in an old cottonwood just below the mouth of Bear Creek. In my lifetime bald eagles faced extinction; now I call in my cats when the great wings beat overhead, when I spot the unmistakable white-beacon head of the predator.

The symbol of America is here in our woods and our meadow, and she has fledglings ready to fly. Bald eagles will only nest in good fishing habitat. Their presence tells me that Bear Creek and the Blackroot River are healing. And next spring tailed frogs will be emerging from the mud of Evelyn Holbrook's pond. They also give me hope.

Evelyn has gone gray like me, and a bit arthritic, and I don't see her riding horseback on the logging roads like she used to, but she and Allen are gathering firewood this fall down in the pumpkin-orange brush along Bear Creek, among stumps and deadfalls, and she calls to me as I walk with my dogs on the logging road above. Evelyn is still a passionate defender of our common stream, our common life.

"You don't own this place," she told the Plum Creek foresters a few years ago, before they logged the Bear Creek bottoms, and the truth of her utterance reverberates in the flowing waters at her feet, where life goes on. "This is God's country," said Evelyn, "and if you destroy it, God help you!"

IN THIS WE ARE NATIVE

One frame passes over the source of light, stops a fraction of a second, and is followed by another. The light is intermittent, the movement jerky. Our minds fill in the blanks, telling us the images are continuous and alive. A movie.

"Only by increasing our powers of vision will we make the death of this world resemble the life of the other [world]" wrote my husband, David James Smith, in 1974, shortly before death removed him from our world at the age of forty-one.

"The experience furnished by the imagination," he said, echoing Wordsworth, "gives us glimpses into life, into eternity, into the world whence we came and to which we go."

I found those words months after Dave died, scribbled on top of a sheaf of notes he'd been working on when his heart stopped beating. He was writing a scenario based on the life of the English poet and artist William Blake, and his imagination had been engaged with Blake's angels and demons and a flea that wears the face of a man. This would be an opera combining poetry, drama, cinema, music, and paintings.

Dave's notes on Blake had been squirreled away for more than twenty years when I decided to read them, the orange plastic crate that held them stacked with boxes of mildewing files in the red-painted shack we call the study. Our four sons were grown by then and had moved from the Montana homestead where their father died, but I stayed in the log house we'd built—alone most days, aging, and, since about 1980, paired with the writer William Kittredge.

Years passed, and still I avoided the cache of Dave's work. I could not bear to see his handwriting—open and boyish, distinctive as his flat, Minnesota voice—or to inflict on myself the disappointment of his unrealized dreams.

I married David James Smith in 1955, when I was nineteen and he was twenty-two, and though we were strong in love we could not have been more different. He was rural, Baptist bred, blue collar, poor, small town, blond, athletic, smart, and funny. He had no father, and his teenage mother took off when he was two years old to start a new life in Chicago, leaving him in Hastings, Minnesota, with his grandmother, six aunts, an uncle, and a slew of cousins.

I was a dark-haired, bookish, city girl, smiling but secretly rebellious, born in Paris and raised by Hungarian Jewish parents. We lived on Chicago's North Side with my mother's mother, who cared for me and my little sisters, Kathy and Carole, while my parents worked downtown as photographers. My father supervised the content of our lives (Mother was in charge of style), leading us toward education, art, music, leftist politics, and travel—a pattern we girls passed on to our kids.

When Dave and I married, he was a law student at the University of Chicago on a basketball ride at a school that had just abolished football. After his graduation, we went to Seattle (a place I had never seen), where Dave practiced law, became disillusioned with it, and got a Ph.D. in English literature. In 1964 we settled in Missoula, Montana, where he taught Romantic poets and the Victorian novel at the university.

Although Dave and I had incurable differences, what kept us together (beyond sexual attraction and children) was a common sensibility about art and ideas, a politics based on racial, gender, and class equality, and a love of the West. We craved adventure and were open to change and risk. Which is why, in all our flirtations with profession and place, I was Dave's partner (although I must admit he was often the leader—probably because his discontent was more deeply seated than mine).

By the time Dave discovered that he suffered from a fatal, inherited metabolic disease that clogged his arteries and endangered his heart, he'd become bored with teaching and obsessed with cinema. He wanted to write and direct movies. Only thirty-five, he was young enough for a third career, but our cardiologist told us he'd never get old. "You might possibly live to be sixty," the doctor said, "or you could die any day."

For five years, we waited. We bought land and built a log house. It wasn't enough to satisfy Dave. He wrote scripts in his off hours, but we were a long way from the source. Dave wanted to go to Hollywood. I wanted him to do what he wanted, as long as we stayed together as a family. My folks and many of our friends thought we were beyond heedless and into crazy. We didn't care what anyone thought. In 1973 he quit his tenured professor's job and we set off for Los Angeles with Steve, who was fifteen, and the six-year-old twins, Alex and Andrew, leaving seventeen-year-old Eric to finish high school and tend the ranch.

We spent half of the last year of Dave's life in a shabby one-bedroom apartment in the Amor Arms on Orchid Street, a block from Hollywood Boulevard. I went there with Alex last spring on a sentimental journey and was astounded to discover that most of our street had become a hole seven stories deep. The slum we'd once shared with the homeless, the hapless, and hopefuls like us was being gentrified. There would be an underground garage, flowered walkways, condos, and malls. But the Amor Arms still stood, a memento of old Hollywood rising at the edge of the pit across from the skeletal frame of the Holiday Inn.

Alex lives in Los Angeles these days, not far from where we lived, but has few memories of the months we spent on Orchid Street. He remembers little-boy things: playing four-square in the asphalt yard of his elementary school; the fat woman who baby-sat him and

Andrew in her apartment off Sunset Boulevard; a handkerchief park (now gone) named Rudolph Valentino, where he tossed a foam-rubber football with his brothers and dad. He barely remembers his father.

David sits in my memory like a bird on a stone. I see him slack shouldered on a dinette chair in the Amor Arms, sallow skinned, eyes dark circled, obsessively writing. Literally locking himself in the bedroom, he writes hundreds of pages longhand on sheaves of yellow-striped legal pads.

"*Innocence,*" he writes, "is distance from death. We aren't innocent, then suddenly experienced." This is the first page of a stack of notes for a script called *Spares.* Not his Blake opera (although the language is undoubtedly Blakean), but a sci-fi movie about illegal trade in spare organs reaped from runaway kids. He is writing about the street culture on Hollywood Boulevard in which teens are expendable and predatory suicide cults can pass for religion.

A few lines down the page he adds: "It looks like I may make a small fortune by Christmas." *What could he have been imagining?* And finally, near the bottom, "—Re 'life': It's only a matter of time (the medical cliché—negative prognosis)."

My skin tingles with guilt when I read Dave's thoughts. They are revelations private as the sound of his blood. Maybe we should not know what those closest to us have written in their journals. Maybe we should bury their notebooks, burn their letters. Yet there is something immediate in the surprise of Dave's words—the scrawled hand, the yellow sheets—and valuable beyond the circle of kin.

"Do you have any recordings of Dad?" Alex asks. He and Andrew have become writers and filmmakers like their parents, but have no clear sense of the father they resemble.

"I'd like to hear his voice . . ."

The snatches of David's voice that I can retrieve, along with his photos and words, are fragments of his being. The story that follows is also a fragment, a fiction created by memory. We cannot reconstruct the dead. David remains the mystery at the heart of our lives—moving, continuous, intermittent, alive.

David and I arrived in Hollywood in the autumn of 1973 with a portable typewriter, two liquor cartons full of books, four suitcases of summer clothes, and a bag full of the twins' toy soldiers and plastic dinosaurs. Our car was a secondhand blue 1965 Ford DeLuxe purchased before we left Missoula for $250.

We were not quite broke, having sold our Land Rover (the long-wheel-base model we had bought in England just after the twins were born) for cash to live on in California. Our Rover was the color of desert sand and we loved it, but it was a lemon and expensive to fix, so we sold it to a real-estate broker for two thousand dollars. Somewhere on I-5 (Bakersfield, I believe) the Ford gasped and huffed. A Mexican mechanic adjusted the carburetor. When we handed him our Exxon card (the only credit card we owned), he returned it sliced diagonally, like toast. It had been denied.

"Sorry," he said, "you'll have to pay cash."

We had come to California for the gold. The green pastures and cows of our Montana homestead were no longer adequate. Our dreams burned with neon light. Both of us were in a hurry, and willing to gamble because a long shot was better than no shot, and we thought we had nothing to lose. We were going to break into *the movies.*

The movies. What a superb American phrase. Takes me back to childhood. I am sitting in the balcony of the Granada Theatre on Clark Street. Shirley Temple is tapping patent-leather shoes. Her knees are dimpled. She is all dimples. My little sister Kathy and I think Shirley is a hypocritical twerp. We have gone to a Sunday double feature with our short and perfectly round Hungarian grandmother. Grandma Deutch gives off an odor of violets mixed with popcorn and old age. The second feature will be *A Night at the Opera.* Grandma loves the Marx Brothers, especially Zeppo (she's the only Zeppo fan I have ever known), and she adores the dreaded Shirley. Kat and I skip home, holding hands, under the happy trance of the movies.

In David's Minnesota Bible Belt childhood movies were forbidden, like dancing and liquor and sex. He told me of sneaking into theaters, the thrill of stage-door exits, velvet curtains heavy with dust. Dave escaped the confines of Hastings right after high school, but he would always be a small-town kid wanting out. Cinema remained strong in his heart, like first love—illicit, and thus everlastingly seductive.

When we came to Montana in 1964, Dave was healthy and energetic and eager to try his hand at making films.

He organized a series—Renoir one Thursday, Godard the next—and finagled the English Department into letting him teach super-eight filmmaking workshops in our Missoula apartment. Our home was furnished with tables and sofas from the Salvation Army store across the street, where we exchanged kids' clothes on a regular basis, but Dave saved enough cash to buy a spring-wound Bolex sixteen-millimeter camera and a Uher tape recorder. We shot a voice-over documentary about our poet friend Richard Hugo, using cheap "ends" of black-and-white negative salvaged from other films. Then Dave experimented with a sync-sound system described in one of the underground film mags we subscribed to, and we set out to make avant-garde low-budget films, like Stan Brakage.

"Recognition and gifts," said Dave, laughing at his desires. It was a catchphrase literally stolen from the Birthday Special advertised on screen at the Fox Theater in Missoula. The bastard boy from Hastings was determined to get his share.

Dave did not turn to writing feature scripts until after his first heart attack. Coming home from the intensive care unit at Saint Patrick's Hospital, he barricaded himself in the second-floor study of our Victorian flat while I tried to keep the squealing one-year-old twins quiet. Surrounded by floor-to-ceiling shelves of the collected books of our lives, listening to Brandenburg concertos or Brahms, dreamy from antidepressants and blood-thinning drugs, David wrote movie stories and planned their production, burying his fear of death until I shouted supper.

He adapted a black-comedy novel about Northwest Coast Indians and, inspired by Sam Peckinpah, researched and wrote a treatment about Pike Landusky—a wolf-hunting outlaw with half his face shot off. After we moved to the Blackfoot valley, nights when we were not building our log house, David and I wrote proposals for historical PBS television series about Nez Perce Indians and Montana vigilantes, and we dreamed up a country-and-western story based on Tammy Wynette and George Jones. Our files included treatments on Red (Indian) Power and criminal justice, and a bunch of ideas I have forgotten. All this is to explain why our decision to go to Hollywood was neither sudden or rash.

Hollywood would be our last frontier, but David and I were not so dumb that we would sell our land. We knew home is the place you come back to (you don't have to live there all the time), and we hoped to come home to Montana within a year with contracts, work, money in the bank.

"It's now or never," Dave said. Common sense could go fly a kite. He had stories to tell, images bluer than Montana skies. One day in summer, before we took off for California, I saw smoke coming out of our cement-block garage. Dave was throwing papers into a fire in a fifty-gallon oil barrel: notes on Dickens and the Romantic poets; scholarly articles; his Ph.D. dissertation on incest in Victorian novels.

I told him to stop. He owed his children an intellectual inheritance. I told him his work was part of their heritage. "It's shit," he said. "My shit. And I can burn it if I want."

"No bridges" was David's war song. Make up your

mind. Walk away, as a friend of mine says, like a movie star. But burning your work can also be an act of bitterness, of self-loathing and disappointment. I wept.

We rented our not-quite-finished log house to graduate students, leaving Eric to live in the bunkhouse and take care of the ranch while he finished high school. Longhaired Steve would be a sophomore at Hollywood High, and Alex and Andrew would enter first grade in Los Angeles.

In Dave's black-leather briefcase (the one I bought for him in 1958, when he went into law practice in Seattle) were names of agents; letters of recommendation from media friends; a telephone number for my father's accountant's son, who was an executive at Warner Brothers TV; treatments and two scripts; unpaid doctors' bills. It wasn't much.

A Santa Ana wind rattled the palms on Sunset Boulevard near the dirty, art-deco stucco of Hollywood High. Inside, Steve was learning about racial wars, hard drugs, and gangs—a tougher scene than he'd experienced in Missoula, where beer, grass and windowpane acid were as bad-ass as a boy like him would get. Had he told us about skipping school and hanging out drunk on Venice Beach, or about passing out from Seconal and Jim Beam with a pal who wanted to commit suicide, or about a friend who'd become a gay prostitute, we would have yanked him out, home-schooled him, sent him back to Montana. At least, in hindsight we should have. But I was so stricken with Dave's problems that I was blind to Steve's.

I shouldered my groceries and walked out of Safeway's cool factory air. It was five o'clock, rush hour at La Brea and Sunset. Rolls-Royces, Mercedes, BMWs, Jaguars surged gleaming on all sides of me. I had never seen so many exotic cars, so clean. The favorite that year was Porsche. Mostly they were leased. Hollywood, we'd discovered, was artifice. You could rent poolside apartments, furniture of glass and chrome, diamonds, designer dresses, everything but talent and luck.

Once, when we still thought luck was coming, Dave and I walked into a Porsche dealer's showroom. A young salesman, dressed for prep school in a navy blazer and silk rep tie, told us the cheapest model cost twelve thousand dollars—big bucks in those days.

"Not good enough," said Dave with the slightest trace of a British accent. "We've decided on a Jaguar XKE."

"What color?" I asked.

"Silver . . . perhaps." We angled across Sunset toward our old blue Ford.

"Yellow," I decided. Our Jaguar would be the color of tropical canaries.

Our one-bedroom furnished apartment in the Amor Arms on Orchid Street sat diagonally across from the back entrance of the Hollywood Holiday Inn and one block from Graumann's Chinese Theatre. It was yellow brick, the manager's rooms on the left side of the narrow hallway as you entered, our two rooms down the hall on the right. Aside from the divorced landlady, her boyfriend, and her baby daughter, we were the only family in a building designed for brief encounters that rented by the day, the week, the month.

The place was cheap and in the center of the Hollywood nostalgia zone, and when we paid our first month's rent we did not know we had landed in a hive of transvestites. Our neighbors came and went in clouds of perfume and hot pants, but purple mascara and pancake makeup could not hide the five o'clock shadows. During our daily walks (Dave was supposed to walk at least an hour a day to develop capillaries around his clogged arteries) we played guessing games to discover who was what. "Watch the hips," I said. "A man's hips will give him away." On Sunset Boulevard sex was as easy to buy as dope or T-shirts. Before a month had passed, we'd quit being starstruck tourists and started being homesick.

Once, a tenant pulled a knife on the landlady's boyfriend in the dim hallway outside our door. We were watching some cop drama on our rented television when the ruckus began. "I peeped through the peephole," Steve remembers. "I saw the landlord point his gun. It was more exciting than *Police Story*." When we went to Seattle for Christmas, someone jimmied the door to our flat, stealing only Dave's prescription heart drugs and his go-to-meetings suit.

Here is a typical scene from those days: I enter the living room, arms loaded with groceries. The twins are throwing toys. Cartoons blare from the TV. Plastic dinosaurs lie beheaded on the fading red shag rug.

"Where's Daddy?" I say. "Clean up this goddamn mess!" I pull Andrew's ear, boot a khaki tank in Alex's direction.

The kitchen is a converted closet, enough room for a hotplate, a mini-refrigerator, a sink, and a linoleum-covered counter. Cooking is simple. We eat stuff I can fry in a skillet, boil in a pot, cut up raw, or pour out of a cardboard box. Tonight it will be canned corn, hot dogs, avocado and grapefruit salad. Fresh fruit and vegetables are a high point of Los Angeles life. The food is miles better than Montana fare, where in 1973 the fanciest restaurant meal was lobsteak—a slab of sirloin bedded next to a defrosted lobster tail.

"I said where the hell is Daddy?"

"Went to the post office," says Alex.

The Hollywood P.O. was where we conducted business. Also the telephone booths in the Holiday Inn. We had rented a post office box because our address was so disreputable. We could not afford the deposit for a phone.

This was our deal. David would write scripts and I would find a production job in what we called "the business." We made a bookcase of pine boards and bricks. By placing our Olivetti on an end table and moving a chair from the dinette, the bedroom doubled as a study. Every morning after I took the twins to first grade at the Gardner Street School, Dave would put on his blue terry-cloth robe, plug in the coffeepot, and disappear into the bedroom, locking the door.

Then I'd be off to the newsstand, where you could browse anything from *Gentleman's Quarterly* to *Professional Soldier*, and read the news in Chinese or Russian. But like dozens of unemployed dreamers, I'd be buying the *L.A. Times* and the trades—*Daily Variety* and the *Hollywood Reporter*—and I'd be checking out the help-wanteds.

The odor of fresh doughnuts seduced me, and I'd return home with a sticky treat. My self-ordained role was to be The Good Mother. I was determined to keep up a pretense of normalcy, but I'm not a good or consistent actress, and more times than not I'd break into true character. I'd shatter the calm by starting an argument.

"I think I understand suicide," Dave said one morning. "We all die sometime. If day in, day out you loathe your life, why not get it over with?

"I wake up in the morning," he said, "and I don't know where I am. I don't *recognize* things. A roomful of somebody else's furniture."

In the notes he had written from those dark times, I found these phrases: "Our life is a trail strewn with pieces of ourselves." "Our life is a trail of strewn pieces of ourselves." "We mark the trail of our life by strewing pieces of ourselves. It's an expensive outing." "We strew pieces of ourselves to mark the trail of our life."

We talked for hours about stories. To add light to the teenage horror show Dave was writing, we created a heroine who saw through the sham, tried to escape, brought down the organ-dealing bad guys. But the horror within our lives remained. We were jammed into a one-bedroom flat decorated like a bordello with red brocade wallpaper and gilt fixtures. Each night we put the twins to sleep in our bed, then transferred them head-to-toe onto the long couch in the living room when we turned in. Steve slept on a foam-rubber mat along the wall by the television. I worried about the twins being molested on the street or in our building (although I sometimes left them with a personable young sailor on the second floor). I worried about the transsexual dancer who practiced his-or-her routine each afternoon in the room above our heads, thumping to melodies we could not hear. I worried about Steve. And Eric living unsupervised on the ranch.

One evening a few weeks after we arrived, LAPD officers stopped Steve in front of our building and frisked him spread-eagled against their squad car.

"They didn't find my switchblade," he told me years later, when he'd become the most law-abiding and righteous member of our family, and we could talk about such things.

"How did you get a switchblade?"

"Think I traded a lid for it."

"How could you afford a lid?"

"I don't know," said Steve. "Sometimes we skipped school, and since you were so broke, I'd panhandle so I could play pool."

Dave and I had come to Hollywood because we believed in the power of place. We came to be inspired. I didn't mind the squalor (we were used to living poor), but we were putting our children in danger.

"We've got to move," I insisted.

"Fine," Dave said. "Go find a job . . . You're just like your mother—bourgeois to the core." He always hit me with that one.

"Shut up about my mother! She cares about her children."

"And I don't care about mine . . . ?"

"I didn't say that."

"The hell you didn't!"

That's how we spent many mornings arguing round

and round about Dave's script, the children, how to salvage our lives.

"Something," said Dave, trying to lighten the situation (mimicking Dickens's Mister Micawber), "will turn up."

So all the months we lived on Orchid Street I was planning to move. The search gave me an excuse to get out of the grim apartment and away from Dave's melancholy. I walked the flat grids from Vermont to Fairfax, from Melrose to Highland. The most FOR RENT signs were tacked up on motel-like buildings. Inside their courtyards were tiled swimming pools edged with slime and soggy green indoor/outdoor carpets.

"Sorry, dear, we don't allow children."

The elderly managers always called me dear. As I walked away, small dogs yapped behind double-locked doors. So I took to the hills. I wandered the canyons of old Hollywood, past cottages smothered in jungle vines, decayed haciendas, Hopalong Cassidy's mansion marked in red in my guidebook map of the stars.

I gave myself to daydreams. Our house would be white. Hummingbirds would flit in bougainvillea draped around French doors like magenta curtains. David would drive off to the studio in his Jaguar XKE, while Rosa, our Mexican maid (whom we were hiding from Immigration agents), took care of the housework and the kids. Everyone gone, I'd sit in my flower-draped patio, my skin bare and gleaming with cocoa butter, a notebook at the foot of my chaise lounge on which to jot lines of poems that settled delicately, in my mind's eye, like South American butterflies.

Perhaps it was the long Montana winters that drove us to plan an escape to southern California. At the ranch, when darkness pulled us indoors at four in the afternoon, the temperature sliding toward ten below, Dave and I used to huddle in front of the woodstove in the bunkhouse and read books about Hollywood in its glory: *Day of the Locust* and *Hollywood Babylon* and *What Makes Sammy Run.* From a snowed-in cabin in Montana, sleaziness seemed glamorous—a film noir starring Bogart and Bacall—but from our vantage point at the Amor Arms, it was obvious that Hollywood Boulevard was the end of the line.

The only way out was to find a job. Phone numbers and résumés in hand, I'd cross Orchid Street and enter the Holiday Inn. Jostling past tourists, I joined the prostitutes and dope dealers at the phone booths—motel regulars who dialed by my side wearing satin hot pants, sequined jeans, and platform shoes. I responded to dozens of ads, but with no connections or experience in the film business I couldn't convince anyone to hire me as a production assistant, script reader, or fledgling agent.

Finally Haskell Wexler agreed to an interview. He was one of my handful of personal connections, having once apprenticed in my father's photographic studio. Wexler was filming commercials at the time, trying to raise money to produce his next feature, which, he claimed, was too political to be financed by a studio.

"Your résumé reads like a novel," he said. I did not know if I should be flattered or embarrassed, but Haskell had no jobs. "Check back" he said. Micawber again. "Something will turn up."

The next nibble I received was from Roger Corman. Corman was our hero of low-budget films, giving young unknowns a chance in his B-movie mill, and Dave was as excited as I at this break. I like to think Corman's people tried to return my calls but couldn't because we had no phone. Then they lost interest. It's hard to accept being so easily disposable. I remember standing down the street from his office building (it had one of those outdoor glass elevators) and wondering if the man in the trench coat going up in the glass box could be him. It was like a grade-school crush, observing the object of desire from afar.

A week or so later I answered an ad at *Playgirl*. They were looking for a literary editor, someone who could call John Updike or Philip Roth or Eudora Welty and say, "Hi, John [or Phil, or Eudora honey] . . . how about a story?"

I'd been an editor at a university press and for a business quarterly, but the writers I called by their first names were too western or old or as yet unknown, and I knew almost no nationally recognized women fiction writers. So *Playgirl* offered me the lower-paying job of editorial assistant.

"Your work," said the tall blond editor from behind her oversized tinted tortoiseshell glasses, "will include interviewing the Playboy-of-the-Month for our centerfold."

I would go with a photographer to the mansion of some sex idol—say Burt Reynolds—and quiz him about his sex life from a woman's point of view. Meanwhile the photographer would be posing Burt in a jockstrap under a palm by the aquamarine kidney-shaped pool.

This was not the way I had imagined cracking Hollywood, but it offered tantalizing possibilities.

"No," said Dave, "it's stupid and demeaning."

"It's a start," I said. "I'll make lots of connections."

"That's what I'm afraid of."

Like many men of his generation Dave wanted freedom for himself but was possessive of me, and like many women raised in the 1950s I was of two minds about sexual liberation. Part of me wanted to be possessed, above all to please. I procrastinated. When I called *Playgirl* back two weeks later, the job was taken.

The other job I refused was for a start-up company with big ideas for subscription television on cable—a new idea called Home Box Office. "You'll advance as the company grows," promised the prime mover. "You'll be on the ground floor." The only opening they could offer me was for a secretary-receptionist.

"I didn't come to Hollywood to type other people's letters," I told Dave. I don't regret saying no to the infant HBO, but sometimes I wonder what my life would have been if I'd taken that job. Would I be rich? Famous? Disenchanted? Someone else?

Sundays we walked the beach in Venice, between Santa Monica and Playa del Rey—the part of the beach set aside for nude bathing. It amused us to come with the kids to watch roller skaters on the boardwalks, bodybuilders, conga drummers with coteries of dancing girls. I remember a lank-haired young man beating out show tunes on a white piano set on wheels. He was better than anyone I'd heard in Montana.

Things had taken a turn for the better. We had gone to the Warner Brothers Studio in Burbank, and my father's accountant's son (an executive in the television division) said he liked Dave's scripts. Since Dave had written a western, he thought he'd be a good bet to write for NBC's hit *Kung Fu* series.

"You can tell it like it is," my family friend told David. Of course, you'll have to stick to our format: a fight here; commercial break there; love interest; free-for-all climax where karate beats the bullets. "Yeah, right," laughed Dave as we crossed the Hollywood divide. "Tell it like it is!"

A television western starring a white guy (David Carradine) playing a Chinese hero in a karate suit was not exactly the break Dave had in mind when he quit his tenured job. To make matters worse, there would be no money up front (beginners write on spec), and he'd be working as a scab (a Writer's Guild strike had paralyzed the industry). But we were at the end of our tether. Writing for *Kung Fu* was a chance to slip inside the locked doors.

I say "we" because I'd be partners on the research and plotting of our first script, and if it proved a winner I'd be writing more. We spent our pleasantest hours digging up stories about Chinese in the downtown branch of the Public Library (a lovely art-deco building, since burned down, with arched doorways, fountains, Mexican tiles). These days you can find a slew of scholarly books about Chinese and Mexicans, Blacks and Women and Native Americans in the West. But in 1973 the few such stories that existed were buried in journals, diaries, and regional archives. They were lost in the blur of cowboy myth that still dominated notions of western history.

We found one typical and yet amazing story about an abandoned Wyoming gold mine. It seems a group of Chinese coolies had sifted through the tailings of the abandoned mine and were gathering leftover gold—enough for a small fortune. The town's white folks got wind of it. The sheriff formed a posse, who forcibly rounded up the Chinese and packed them into a boxcar.

"We're sending you home to China," explained the sheriff. Then the boxcar was turned loose on its track down a steep grade into a boulder-filled gully. After a few rounds of whiskey, the posse rode into the gully. The boxcar was not smashed to bits. It stood intact among the boulders, except for the doors, which were open. Not a single Chinese coolie could be seen.

We did not know how to fit such a story into the *Kung Fu* formula. None of the actual stories fit. We lost enthusiasm and hope. "I don't have time for this," said Dave. I agreed. He had come to work on stories that were compelling. So we abandoned our best chance for success and went back to unemployment and walking the streets.

The place we walked most was Hollywood Boulevard. Walking was good medicine for Dave's heart and my soul. Walking eased tension, helped us think more clearly, and put us in touch with a world beyond our immediate problems. On sidewalks embedded with five-pointed stars, each star holding a golden name—Judy Garland, John Wayne, Groucho Marx—we rubbed shoulders with tourists, bums and bag ladies, child prostitutes, heroin dealers, exiles from Cuba and El Salvador, and the beautiful silken transvestites. I remember a homeless guy maybe six foot seven, an ex-army sergeant in a khaki greatcoat and size-thirteen sneakers. We passed him almost every day, and every day he shifted his cardboard suitcase from his right hand to his left and touched the rim of his World War II cap to greet us.

By March, the days were sweet in the Hollywood Hills. The evening sun cast a pink glow on Marilyn Monroe's red candy lips in the windows of the Wax Museum. We passed a bag lady we'd seen many times—the one with a runt dog dressed in a green sweater. The dog stood on the baby seat of the lady's shopping cart, which had been stolen from Safeway. She smiled at us like an old pal.

I was in no mood for smiling. Dave had just gotten out of Hollywood Presbyterian Hospital. We thought he'd suffered another heart attack, but this attack had been gallstones. Painful, but curable, and scary as hell. He was pale and shaky. Not only that, but our friendly transvestite neighbors moved out of the Amor Arms and were replaced by pimps and teenage prostitutes. The pimps were black, cool, and dangerous, breaking in the girls. The girls acted like high-school cheerleaders. They were slim and blond, and they giggled.

"I want to go home," said Dave.

It was my turn to resist. I had gotten used to Hollywood. I wanted to live in Santa Monica, looking out to an infinity of water and sky and sun-filled days. Dave had finished his notes and was beginning to write the script called *Spares*. I was confident we could sell it. And I had become addicted to the freeways. Like Joan Didion's Maria in *Play It As It Lays*, I played the freeway game, wheeling our Ford DeLuxe up ramps and off without missing a beat or letting anyone pass. I'd drive from the Hollywood Freeway to the San Diego, then over to the Valley and on to the Pasadena, timing my return to West Hollywood to pick up the twins after school.

I even had a lead on a job. A lawyer friend represented Lily Tomlin. Her manager needed an associate. He had recommended me. Finally, I thought, we might connect with the talent and wit that drew us to Hollywood. It takes time to get started, I said, stating the obvious. I said we should wait until summer, argued that the kids should finish their school year. Where will we stay? I asked. Our renters in Montana had signed a lease that ran until June.

"I can't wait," said Dave. "We'll break the lease." There was no arguing with him. "I can write at home just as well as in this hellhole."

Where I was slow to make up my mind, but adaptable and tenacious—a player who hangs on at all costs—Dave was decisive and authoritarian. He followed his instincts and desires, being in love with creation not consistency, and failure didn't frighten him as it did me. I'm not sure what triggered his sudden decision. Perhaps he sensed an imminent threat to his health. Or spring's homing instinct took him like it takes monarch butterflies and Pacific salmon—magnetic forces pulling native species back to primal grounds.

We packed the portable Olivetti. The xeroxed scripts and treatments and notes. The unpaid doctors' bills. We packed our four suitcases of summer clothes, our liquor boxes full of books, our cameras and sacks of dinosaurs, little armymen, crayons and coloring books. Alex and Andrew leaped from box to box in the living room. They were cowboys. They were superheroes.

"Can we get a new kitty?" During one of our weekly calls to Eric back in Montana, he'd informed us that the twins' cats had died of distemper during the winter.

Sure, we promised. They could get kittens right away. As we drove east out of Los Angeles on the freeway, the loaded Ford blew a tire. Our spare was flat, such was our luck. I held tight to the twins while semis roared past, creating noxious winds that could knock a child to the ground. Dave and Steve rolled the tire down an off ramp to a garage. I felt like Ma Joad, itinerant, immigrant, stranded, and broke.

California disappeared in our rearview mirror. We switched the radio from pop music to country and western. We sang, *All the gold in California / is in a bank in the middle of Beverly Hills / in somebody else's name.* We sang, *California's a garden of Eden / a paradise to live in and see / but believe it or not, you won't find it so hot / if you don't got the Do-Re-Mi.* When we topped the Sierras and headed into the true desert West, we sang, *This land is your land, this land is my land.* And when we crossed the Columbia at Vantage in Washington State, we sang Woody Guthrie's *Roll on Columbia, roll on.*

"We're home, boys," said Dave as we topped Lookout Pass from Idaho, winding down into Montana through deep, melting snows. This was our Rocky Mountain springtime. We talked about the wildflowers we'd find on our meadow: yellow buttercups in the wake of snow, and glacier lilies called dogtooth violets, and purple shooting stars like tiny birds of paradise, and the white-petaled, pink-centered spring beauties the elk loved to eat.

Dave was rereading Blake's *Songs of Innocence* and *Songs of Experience.* He recited the poem about the rose:

Oh, Rose, thou art sick!
The invisible worm
That flies in the night,
In the howling storm,

Has found out thy bed
Of crimson joy,
And his dark secret love,
Does thy life destroy.

This was *his* song. "You could put that poem to music," Dave said. His eyes became animated. His face looked years younger. "You could make a musical with great songs about Blake's life . . . The guy was a revolutionary, a madman, a lover of women." The excitement of ideas rang in his voice, the passion that made him a good teacher. It could be a movie and an opera, or a movie inside an opera, illustrated by Blake's art.

Later he showed me more hopeful lines from the epic poem called *Europe.* In a prefatory Prophecy, the poet asks: "Then Tell me, what is the material world, and is it dead?" A Fairy (the muse who will dictate the poem) replies:

> I will write a book on leaves of flowers . . .
> I'll sing to you . . . and shew you all alive
> The world, where every particle . . . breathes forth its joy.

This is the story all writers want to write. *I can see it,* Dave said as we headed up the Blackfoot valley on the last leg of our journey. At least that's what I imagine him saying.

Memory and imagination tend to merge after a while, and now I remember (imagine) him driving our old Ford up Bear Creek Road, snow dripping in sunshine from firs and ponderosa pines. He is explicating one of his favorite passages from *The Marriage of Heaven and Hell*—words I would find underlined in his dog-eared Oxford edition of Blake: *If the doors of perception were cleansed everything would appear to man as it is, infinite.*

It was the end of March, time for renewal. Dave was beginning to see the work that lay ahead, a conversation with the imagination. He could not know that before wild roses came into bloom on our ditchbanks, his unfinished vision would be packed away in the orange plastic crate he used as a file (nothing left but crimson joy).

Leaving Hollywood like dogs on the run had been

the right thing to do. Dave was home, where he wanted to be. I remember him striding up the meadow in his long-legged way through blue-tufted waves of camas. He carries a shovel and a yellow plastic irrigation dam rolled around a two-by-four. The new grass is velvet green and water spills from the ditch he has dammed with a spadeful of heavy sod. He is holding *Infinity* in one hand, *Eternity* in the other.

TO DIE IN TIJUANA

The day we opened Flossie's letter we were walking in Bel Air where the rich people live. Dave and I often came to Bel Air in the middle of the day when the boys were in school. It was soothing to walk the hushed streets, miles from our hundred-dollar-a-month apartment where even the wallpaper was an artifact of failure.

This was in the fall of 1973, before urban jogging and race-walking became fashionable, so the only other humans we saw on foot were nannies wheeling babies, Mexican servants waiting for a bus, and Japanese gardeners. The gardeners seemed as out of place as we felt. Their battered pickups stood like poor relatives in millionaire streets, loaded with branches of fig and palm, the pruned limbs of lemon trees, dead leaves, and rot. The gardeners wore coveralls and baseball caps. In the oppressive coastal heat their arms glistened with sweat as they planted and raked and trimmed hedges thick with purple blossoms whose names we did not know.

To be a gardener in Los Angeles, I thought, was to be an artist—a sculptor of palms, a painter whose palette was bougainvillea. Oranges would drop heavy and sweet into your hands. Compared to our garden in Montana, where most fruit is stunted and frost kills even in July, this was Eden. Evenings, driving freeways under a red-smog sunset, if I were a gardener, I'd be tired but guiltless. The honest labor of gardening is better for body and soul, I told Dave, than being down and out in Hollywood, trying to break into the movies.

"I'm sick of honest labor," Dave laughed. He was only half joking. The idea of laboring in the fields, he said, is more attractive than the practice of it. Ask any cherry picker or Nebraska farmer retired to Arizona. Ask your own kids how much they like bucking bales for a few bucks a day.

The blue envelope bore Flossie's meticulous round letters, but its stamp was exotic. The return address was not from her home in Wisconsin, but Tijuana.

"You'll be sad to know," the letter began, "that your Aunt Flossie has had to go to Mexico to get cured of her cancer. The tumor's big as a grapefruit in my right lung, and my doctor back home says it's too late to do anything but ease the pain."

Flossie's doctor had scoffed at laetrile, the faddish cure that would take her to Tijuana, calling it Mexican voodoo. The reason it was illegal in the States, she said, was that the AMA hated anything natural. "Laetrile is a *natural* miracle," she wrote, "extracted from the apricot pits." She'd read about it in *Prevention Magazine*, which was the only word Flossie and her next youngest sister Grace would trust outside the Holy Scriptures. Flossie promised to send Dave a subscription (the aunties knew of his heart disease), so he could save his life too. We never received the subscription.

A big woman, rawboned and heavy with tightly curled gray hair that she sometimes wore in a net, Flossie was the bedrock of David's family, the oldest and most severe of his six aunts and one uncle. The Smith sisters had scattered with husbands and children to small towns and farms in Minnesota, Wisconsin, Michigan, and Illinois, but the energy of the family revolved around Flossie. She was at the center of every feud, sometimes not talking to one or more of her sisters for years.

Flossie had spent most of her long life working the dark soil on both sides of the Mississippi in Minnesota and Wisconsin. She was a homebody, not a traveler. We could not imagine her in a desert border town struck down by cancer.

"I have sold my land," said Flossie's letter. It was not a thing to cry over. Her husband, Art, had died a few years back and working the farm was too exhausting. It's time, she said, to lie down and let the world run on its own. Part of the farm money would go to cure her cancer. With the rest, she and Grace (newly widowed and advertising for a mate in the personals of *Prevention Magazine*) planned to retire by the sea in San Diego.

When she married Art, Flossie'd been old for a country bride—almost forty and probably a virgin. They never had children. I used to wonder what their intimate life could be: thin, gentle Art who loved to raise pigs; Flossie strong, puritanical, and a born-again Christian. Dave became their ready-made boy when his mother, Virtue, the youngest of the family and its black sheep, went away to Chicago when he was two.

With long, honey-colored curls and deep blue eyes, Virtue had been the most beautiful of the sisters, and the most sensuous. She was seduced at fifteen by a married Swedish farmer named Almquist, whose family went to her church. Almquist abandoned Virtue, never acknowledging David, although they came to know each other by sight, passing in the streets or at the grocery store. Dave said he was not ashamed for himself ("it wasn't my fault"), but he must have felt anger toward the man who'd made his life so different from boys with respectable families, and he must have felt shame for his mother. In the small community of Hastings, Virtue was an outcast. The stress was severe. She suffered a nervous breakdown.

"You know, Virtue'd always been tidy," Flossie told me once, "she was a good worker—and smart. Our baby. Everybody loved her. But after Dave was born, she got all sloppy. Didn't take care of herself . . . or nothing. It was such a waste!"

The one thing Virtue did take care of was Dave. A doting mother, she nursed him until he was two and gave him all the hugs and kisses a boy could want. But Hastings in the 1930s was a dead end for an unwed mother, and Virtue needed to reclaim her life. So she left Dave while she went away to Moody Bible School, settling in Chicago. She promised to send for the child when she had a steady job and a good home.

David was a charming toddler with flaxen hair and wide blue eyes. Flossie wanted him and convinced Virtue to let her take him. She tended the boy with the same stern care that ruled every living thing in her domain: the black-and-white Holstein cow; her prize bantam chickens; and a large garden of squash and potatoes, peas, beans, sweet corn, cabbages, onions, and carrots. For sweetness, there were strawberry and raspberry patches and rows of luscious tomatoes. Dave remembered hiding a saltshaker among Flossie's tomatoes so he could sneak one, sun warmed and ripe, and devour it on the moist earth of an August morning, his mouth salty and running with juice.

Beyond the garden were rolling fields of oats, barley, corn, and alfalfa. The land sloped toward the Mississippi River with Wisconsin on one side, Minnesota on the other, and the breezy spaces between the banks seemed endless to David, who had no children to play with except after church on Sundays, when some of his cousins might visit. But there was a white-muzzled sheepdog called Shep. And Art's pigs.

Dave loved to tell stories about winter afternoons when he and Uncle Art escaped to the hog shed, where even Flossie could not create order out of wallowing chaos. The rough enclosure was low to the ground and patched with tarpaper. It steamed with warmth and the sour stench of pig shit. Dave kept a pet there, a little pig who learned to come when he was called. As Art said, "Pigs is as smart as dogs."

When Dave was old enough to start school, Flossie and Art wanted to adopt him and keep him on the farm, but Virtue said no. He was carted back to Hastings, where he lived with his grandmother until eighth grade, when she died. Then he stayed with his aunts Mary and Margaret, working in Grace's greenhouse, being an honor-roll student and a guard on his high school's all-state champion basketball team. When Dave won a basketball scholarship to the University of Chicago, he went there instead of to the University of Minnesota because it was near his mother—a decision that changed both of our lives.

Each year Dave grew further in mind and spirit from Aunt Flossie, but he'd imprinted on her like a gosling to its mother goose. When we went looking for land in Montana, it wasn't chance but a homing impulse that led us to buy a homestead ranch with rolling meadows, a log barn, and a pigpen. The first crop Dave and I planted in our newly turned clay soil was an ill-fated row of tomatoes, but I put my foot down at pigs.

Uncle Art grew thinner and more silent each time I saw him until one winter he disappeared with the snow. We had moved to Montana, and it was spring before we heard he was dead. I don't know what took him. Maybe pneumonia. Art's passing marked the beginning of the end for Flossie, and for a way of living I could never fully comprehend. No matter how much I threw myself into country life, I would always be an outsider—an urban immigrant with no roots in the deep-soil culture of farming people.

Once there had been great deciduous forests along the northern reaches of the Mississippi—Potowatami and Chippewa and other woodland tribes hunting passenger pigeons, elk, bison, and wolves. Then came a swarm of pioneer farmers like Art's people and Dave's grandfather (a white-bearded yeoman named Fernando Mercu Smith), who turned up the sod from Minnesota to Nebraska to Montana, logged the forests, and planted wild meadows with corn and wheat. Finally there was only Flossie and her rooting sows, her tomatoes and hollyhocks and unnerving faith. And then she too was gone.

The letter ended with a postscript: "P.S. David, since you are my neighbor in southern California, you might think about being a Good Samaritan and come to visit me."

"We better go soon," said Dave. "If Flossie doesn't expire from the cancer, she'll surely die of loneliness."

We paid twenty-four dollars for round-trip Greyhound tickets to San Diego. I had arranged for the landlady to baby-sit Alex and Andrew and to keep her eye on Steve. A sea breeze had blown away the smog and it was a shining morning when you could imagine California as the first Spaniards saw it—the chaparral hills running seaward below a wall of snow-beaded mountains. There were dolphins at play in the teal-blue surf, and the beaches we glimpsed were spattered with sun and alive with birds. We were wheeling south along the Pacific toward a foreign country, and I was giddy with freedom. I reached for Dave's hand. No matter how grim the reason, we had embarked once more on an adventure.

Our pleasure subsided when Dave and I got off the bus, passports in hand, and walked over the bridge from the United States to Tijuana. This was our first visit to Mexico and it was nothing like passing into Canada or crossing the border from France to Spain. The dry riverbed below was littered with beer cans, tequila bottles, yellow dogs, and excrement. The border had spawned an indigent city made of tarpaper, rusted tin siding, cardboard boxes, and displaced people ghostlike in their poverty.

Barefoot children pounced on American cars, wiped at windows with dirty rags, then demanded payment. One small limping boy with sores on his face and a bandaged hand serenaded us in a high, clear voice, "South of the border, down Mexico way . . ." Dave gave him a quarter.

I was amazed by the crowds, only a few of them tourists like us. Midmorning, midweek, and none of them working. "What do they do all day?"

Dave shrugged. "Same as we do, I guess."

The main street was all temptation—shops and warrens and peddlers. I fingered a soft leather jacket. The peddler lowered his price with every breath. I longed for silver bracelets, the embroidered blouses from Oaxaca, but Dave pulled me away.

"We'll come back when we have more time and money." Time and money. We never would have either.

Streetside taco stands sold sweet-smelling blue corn tortillas, four for a dollar, and forgetting caution, we ate in the sun. We bought a street map and plotted what looked to be a scenic route to Flossie's clinic. A honking trail of taxis followed after us, drivers shouting promises of fun with *putas* and donkeys and sixteen ways to eat a pussy.

We walked uphill past the jai alai amphitheater and billboards touting greyhound races. As the tourist strip dropped away, we found ourselves in a regular city. Children in starched blue-and-white uniforms were coming home from school. Housewives moved in and out of storefront shops crammed with hardware and shoes and dresses you'd see in Kmart. Yes, this was Mexico. I was walking the cracked pavement. I was smelling frying chorizo from open kitchen windows. The lilt of Spanish sang to me from shuttered adobe walls painted pink and aqua and faded orchid. Dust and the blazing sun caused rivulets to run down my neck, my armpits, the clammy insides of my thighs.

Hunger caught up with us, and thirst. We bought lemon sodas in a small grocery, and a sack of sugary pastries from a *paneria* where fresh loaves of bread, round ones and long ones and crusty brown buns, some still warm from the oven, gave off such an odor my mouth filled with saliva. The pastries were cheap and crumbled in your hand. They left a residue of cinnamon on lips and fingers and the lining of your throat. I looked at Dave in his sweat-soaked shirt, wondering if he felt as happy as I did or if he was worrying about his aunt.

"How about roses?" asked Dave. "I'll bet no one's thought to send Flossie some roses."

The florist did not speak English. She wanted to sell us cactus in painted Mexican pots. I had to take her by the hand and lead her to a glassed-in case where a meager assortment of long-stemmed roses drooped in a ceramic urn. We picked out six of the best. Deep red. Expensive, even in pesos. Roses clenched like fists, their silky petals, their perfume.

The one-story brick clinic was unpretentious and surprisingly clean. Its nurses were polite and attentive, and the doctor spoke a deeply accented but comprehensible English. He told Dave that Flossie was "a very strong woman"—a fact that Dave already knew.

The pale green corridor smelled of antiseptic, but under that astringent I sniffed another, incongruous odor. The hair on my forearms rose, as if to warn me of danger. Cabbage! The scent transported me (as only smell can) to my grandmother's kitchen in Chicago. I sniffed the Hungarian stuffed cabbage she made for my father; and the fragrant cabbage strudel laced with apple and raisins; and the rich steam of old-country sweet-and-sour borscht, dark red, topped with a dollop of sour cream. Cabbage, I thought, is like death. Cheap and universal.

A swarthy young man on crutches, one leg amputated at the knee, stumped toward a dining room with white cloths on the tables. An American woman, her red hair tied in a ponytail, waved. "Hi!" she said, recognizing us as compatriots. "This place is great. It's curing me!"

We did not stop to chat, for we were skeptics. What could cure the incurable? The first apartment that Dave and I had shared was married-student housing at the University of Chicago. We lived on Ellis Avenue across from a Victorian brick mansion called Home for the Incurably Ill. It was a blunt and ugly name that we stared at every day, a memento mori for newlyweds in an era that did not believe in euphemisms.

"Do not be surprised," said the doctor, touching Dave's elbow. He tapped at a green metal door. He shrugged. "We do the best we can."

Flossie lay propped by pillows in a hospital bed. We'd roused her out of a nap. "Bad boy!" she exclaimed in a voice more shrill than the one I recalled. "You should've let me know you were coming."

She held out her arms to Dave, and then to me. Bony arms, where flesh flapped from her old biceps, loose, white, and useless and frightening. I was prepared for changes, but not this skeletal old woman whose prominent nose stood out from sunken cheeks. This Flossie reminded me of a bird—some ancient long-beaked species like a pileated woodpecker.

"Surely you know better than to drop in on a person. Don't even have time to comb my hair, powder my nose!"

Flossie in makeup was a sight I could barely imagine. Perhaps she'd fallen to powder puffs in old age, but I doubted it. Dave had taken me to her farm before we were married to get her approval. I'd grown up in Chicago and felt compelled to impress his rural relatives with my hip, city ways, which in 1955 included black turtleneck sweaters, jeans, sandals, and dark-penciled eyes (exactly what I wear today, except the pants are no longer size ten and the eyes are ringed with crow's feet).

Flossie's glance made me feel naked. I remember her saying something about God, and how we don't need to improve on His work. I excused myself to the bathroom and scrubbed my face in hard well water, cloudy with rust and smelling faintly of sulfur. Now Dave was having a hard time keeping his face in order. He fumbled with the brown-paper wrapping that held our gift of flowers. I could tell he was grateful to have something to do with his hands.

"Oh, David!" Flossie cradled the thorny blossoms, rocked back and forth, put on her gold-rimmed spectacles to examine them more closely. "Your Uncle Art was the only man in the world ever thought to give me roses . . ."

Flossie buried her head in the dark petals. We could hear her inhaling. I had to look away. The room was impersonal, bare, and blessedly cool. A breeze sighed through white-slatted venetian blinds. On the wall facing Flossie a painting of Jesus Christ held out an open hand whose stigmata formed a glowing cross. He was golden bearded, blue eyed, smiling. I wondered if this was her own picture, salvaged from the Wisconsin house.

When Flossie lifted her head, tears blurred the sharp glance of her eyes, yellowed with age and enlarged like an owl's behind thick lenses. Dave stood awkwardly by her bed. I could hear his nervous cough as I searched for a vessel in which to put the roses. I found nothing. Flossie had come to Tijuana clean, with no mementos except that icon of Christ and a small framed photograph of Uncle Art on a tractor, the farmhouse in the background tilted and slightly out of focus.

"Maybe I can find a vase," I faltered, and fled from the room, leaving Dave with his flesh and blood.

For Dave, Flossie had been the cast-iron backbone of order and right thinking. Before he was out of diapers, she'd implanted the fear of God in him, and he'd never shake off the burn of her hand on his bare buttocks. One summer day (one of his earliest memories) she found him in the marble birdbath, naked as Adam and pissing with a three-year-old's erection.

"Shame," she said. "Shame, shame, shame." Then she closed him in her closet, where he almost fainted from heat and the poison of mothballs. After that, Dave was uneasy being naked in daylight (night was a different matter). He wore swimming trunks in hot tubs and disdained skinny-dipping. I tried to coax him out of such modesty, but as long as the sun stayed in the sky, he kept his pants securely fastened.

When I returned with a vase, Flossie was nibbling at one of the Mexican cookies we had brought. For me there was little to fear in her except judgment, which is fearful enough for a girl who grew up with no possibility of family rejection. Acceptance was the rule in our tight three-sister circle. My mother might nag, my father might argue, my grandparents might cajole and bribe, but I grew up sure that no matter what I did, I would always be loved.

Not so with David. He was born into rejection, grew up in a culture based on authority and punishment, and was taught Flossie's code of stoic forbearance. It was possible for Dave to reject love. Possible for him to walk away. These were qualities I both feared and envied.

I envied the will that kept Flossie upright, with no mention of disease or pain. When I die, I hope to go fast, taking my chances with heaven or hell. I am not a coward, but lingering, incurable pain might make me one. I might grow irritable and prone to tears like my father did; or rage about injustice and ruination; or retreat into memory and fantasy and drugs.

David was different. He studied his pain, teased it and scratched to see how it worked. He tried to make it into something better. Sometimes depression paralyzed him, but lately he'd been teaching himself to look into his fright, hoping to create art from it. Dave taught me that it is possible (actually, necessary) to transform experience and thus transform yourself, even if you know the job will never be finished.

"How'd you kids know I was dying for sweets?" Flossie brushed crumbs off the rough sheet. "All they serve us is boiled potatoes and cabbage soup. They say it's got vitamins, but I wouldn't feed that slop to my pigs." Flossie's voice rose to its old fierceness. "They're cheapskates—that's what they are!"

She pointed to the remains of her once-proud stomach under the loose cover of her hospital gown. "Look at me! They're trying to starve me to death!"

We were interrupted by a quick tapping. A squat, heavy nurse entered the room. She was holding a cloth-covered tray, which I thought might be Flossie's dinner. But under the cloth were plastic syringes, glass vials, and long shiny needles.

"You better go," Flossie said. "It's not much fun." She lay back on her pillows, swallowed a last bit of cookie. "You'll be coming back?"

"Yes," said David.

"Good," said Flossie. Then she closed her eyes.

It was near dusk on the streets, the October sun glowed low in the sky, time for dinner, but we were not hungry. Mariachi music flowed through the open doors of dark cantinas. I wanted to go inside, to drink margaritas made with Sonoran tequila and fresh-squeezed limes. I wanted to get high with David, to be happy, even to dance. Mostly I wanted us to be lighthearted—a way we hadn't been for five years, not since he got sick with the heart disease that was as incurable as Flossie's cancer.

You could never tell about David. Sober, he might disdain frivolity, bemoan a drinker's loss of control, be puritanical as the teetotalers who'd ruled his youth. A few drinks changed him. Inhibitions gone, he turned charming and joking, or raunchy, or angry as a rattlesnake. Too much booze and he was sick as any human I've ever seen (once he spent two days on his knees with his head in the toilet).

I blamed Dave's dark side—his anger, his sickness, his lust and rebellion, his longing always for something better—on Caleb Almquist, the farmer who had fathered a son and then denied him. I also blamed the small town he grew up in, where busybodies snickered behind the bright boy's back and everyone knew his family's secrets. Heredity or environment. Either way led straight to trouble.

Dave could have known his father as more than a passing face. He went to the same Baptist church; attended high school with his younger, legitimate, identical-twin half sisters. He could have confronted Almquist, but as far as I know father and son never spoke. Years later, returning to Hastings, Virtue became obsessed with trying to arrange a meeting between Dave and his father, but Dave would have none of it.

"Why don't you try it? Do it for your mother," I'd say. "And for us. He can get to know his grandchildren." I have always believed in reconciliation.

In one of my lost-father-come-back fantasies, Eric and Steve toss a Frisbee and the twins play with toy tractors in the dirt. We gather around a picnic table in a park next to a baseball field. There is not much talk, and no recrimination. I bring cake, a thermos filled with hot coffee, a pitcher of ice-cold lemonade. When afternoon shadows reach down from the stands, the old man shakes hands with David in a formal, old-country way.

Dave dismissed my fantasy as he must have dismissed countless ones of his own. But he could not dismiss a likeness that was at least skin deep. One of the photos Dave carried in a cardboar box holding snapshots from his childhood is a portrait of Almquist when he was somewhere near forty. He has a high forehead like Dave, and fair, receding hair. He wears wire-rimmed glasses and looks straight at the camera—a tight-lipped Swede not as handsome as his son. The photo is black and white, but I imagine Caleb's eyes to be blue, like Dave's, although bereft of sparkle. I see a meanness in the father's mouth, rigidity in his posture. I want to know if the mysteries under my husband's skin (genes transmitted to our children, and their children) can be solved by knowing this lost father's story.

Dave didn't want to think about such things. He would not put himself in a position to be rejected again. You have to rise above genetic inheritance, he said, the usual expectations. When his mother insisted they initiate contact, he got angry. "No," he almost shouted. "You can't undo the past."

We killed time in Tijuana like we'd done for nearly twenty years during aimless nights in the distant seaports we'd gone to—Seattle, Oslo, Bergen, Copenhagen, New York, London, Paris, Malaga, San Francisco, Hollywood—walking the avenues, watching the sun disappear. We did not eat in the local cafés that night or drink in the bars. I can't remember what we talked about, but I'm sure our subjects included memory and movies and death. A slow hour passed. We turned back toward the clinic.

Our concern was Flossie's treatment. Dave had read up on laetrile in the Hollywood Public Library and discovered the treatment was extremely painful as well as probably futile. When I asked whether it wouldn't be better to tell Flossie the truth, Dave laughed.

"It doesn't make sense," I said. "To die in Tijuana, when she'd be so much more comfortable at home."

"Believers," Dave said. Flossie was a true believer. She had taught him the comfort of faith and he'd rejected it. Now that he had to face his own dying, I wondered if he regretted the choice he'd made as a young intellectual set free in Chicago, life opening like the Great Lakes, nothing but blue skies at the horizon. He was in his lecture mode. "If you were her and you were dying, wouldn't you grab on to whatever gives you hope?"

I had been taught that science was reason and had little to do with dreams or hope. I believed in reason, but lived in dreams and acted on faith. "I'm talking about truth," I replied.

"Take a load off your feet," said Flossie, patting the side of her bed. Her eyes were dark circled and her voice seemed a bit unsteady. Dave followed her orders and she took hold of his hand. "Do you know about laetrile?"

Dave shook his head in a gestural white lie.

Flossie glared as she talked. "I told you it's made from apricot pits. There's arsenic in them. It'll poison you if you eat'em raw, but these scientists here fix it up so it only poisons the cancer.

"They give me the injections," said Flossie, "here," showing us the needle marks on her arms, "and here, where the cancer is." She pointed to a bandage across her chest, where her heavy breasts hung flat and long under the thin nightgown.

Dave caressed Flossie's hand. He smoothed the soft, wrinkled skin, the faint freckles on her long fingers, the palm that would see no more calluses.

"The tumor's shrinking. I can feel it, you know . . . shrinking." Flossie's eyes seemed to be peering inward. Her voice lowered to nearly a whisper. "Funny how a person gets to feeling what's going on in your insides same as what's happening outside your skin." She aimed her words at me, as if this were the talk of women, talk Dave could not comprehend.

"Never had time or interest in such things before," she said. "There was always work . . . and the church . . .

"You sit quiet and concentrate," said Flossie, "and you feel things moving you never saw before."

"Yes," I smiled, anxious to let Flossie know I shared her knowledge. "That's how I felt when I was pregnant. It wasn't just the kicking and the squirming . . . I could feel those babies growing inside me."

David dropped Flossie's hand. Here was this barren

old woman growing a death-sized tumor, and here was his wife saying it was just like having babies.

"I know what you mean," she said. Her right hand moved to touch the bandage on her breast.

David looked away. Women were sharing an understanding he might never have come to if he had not gotten sick, but he was more tuned in to his insides than I was to mine. I preferred observing birds and people—the weather. Introspection still makes me squeamish. Dave thrived on it. In intensive care he had learned to listen to his heart's beat. He could feel the flow of his blood; note irregularities; measure breath.

All her life, Flossie was saying, she'd seen animals born and slaughtered. She'd pulled calves and lambs, fed the orphans with a bottle, named them and put them out to pasture knowing they would be sold and butchered in the fall. "I never liked the butchering," she said. "Not even the chickens."

"Yeah," said Dave, back in the conversation. Flossie let out a long sigh.

"Pigs was worse," she said.

"That's because they're more like humans," said Dave.

I remembered the birth of our twins, so pink and tiny and helpless. How I'd felt like a bitch suckling her litter.

Flossie cocked her head. "I've seen life at both ends," she said, "especially the hind end." Her eyes were impertinent. With her head at that angle, she reminded me of a raven. Its mocking stance. Its inherent blackness.

Dave threw back his head and laughed. At times, his laughter could be mocking, hard hearted, but Flossie knew this was not such a time. She laughed too, holding her arms around her chest to muffle the pain. We all laughed. Tears gathered in David's eyes. It was rain when you needed it.

Aunt Flossie never got out of Tijuana, and we never went back. Our troubles became overwhelming and we left Los Angeles at the end of March, arriving in Montana with the robins and the bluebirds. It was runoff time at the ranch, new grass coming green at the edge of snow, the ditches flooding, water racing down from the pines and across our meadow in swaths of silver. Flossie was dead by then.

I guess Grace came to claim her, or one of the younger sisters: Beulah or Mae Belle, Mary or Margaret. I'm sure her remains were shipped back to Wisconsin, for she wanted to be buried next to Art in their plot at the cemetery near the farm.

"I've got my six feet bought and paid for," Flossie told us as we said our good-byes in the Tijuana clinic. Her name, she said, was engraved on the double tombstone. "When my time comes, all you got to do is fill in the date."

Dave died too that spring and we buried him in the old Catholic cemetery in Missoula, on a tree-shaded slope under the freeway and overlooking the Northside softball fields. Dave had loved to play ball and his boys would follow his tradition when they graduated to men's league, playing on that field where he might see them, on a team called the Montana Review of Books.

I bring wildflowers to his grave each year, and half a dozen long-stemmed roses—deep red like the ones he gave Flossie—the six roses signifying our family, like our brand 6s, five for me and the boys, one for him. I place them on his black marble gravestone, near the white engraved lines borrowed from W. H. Auden: WHAT THOU LOVEST WELL REMAINS.

In my mind I also bring roses to Flossie's grave. I've never made the pilgrimage, but I imagine a green cemetery shaded with scarlet-leaved maples. It looks down to the Mississippi River over farms whose white houses are backed up to gardens. The gardens are bright with hollyhocks, the tomatoes in them are red and ripe. From the steepled church, a bell rings as if from an earlier century in a village in Germany or Sweden. Snow geese rise in thousands from the glinting river, a snowstorm in reverse. I hear the beating of enormous wings.

I kissed Flossie before we walked away that October evening long ago—before Dave and I crossed the Mexican border back to a land where we could, at least, speak the language. Her breath smelled of rotten apples, and her lips were dry as rice paper.

David and I wept for Flossie as we walked over the bridge from Tijuana to San Diego. In the chasm under the bridge, darkness hid the garbage; bonfires glowed like fallen stars; dogs barked; a mother called to her child in Spanish. We passed through customs, and then we got on the bus. We did not know that we were weeping for ourselves.

THE WIDOW WOMAN

A day will come that changes your life. It will be an unpremeditated afternoon, blue skies or rain, nothing like the sacramental days of marriage, birth, and death. It will catch you unawares like a night etched with betrayal, or first love, or spiritual ecstasy, or a morning in the Grand Canyon when fog lifts to light on red rock, the river molten, and you are felled as if by sunstroke with the immense, surprising beauty.

What I'm talking about is a telephone call one Tuesday in July, Walt Schaar, the station manager for KSPS public television in Spokane, calling to ask Dave to be the executive producer for a series of programs about seven tribes in the inland Northwest—a two-year job—and asking me to be associate producer. Good salaries. Benefits. A socially responsible media venture with creative freedom. Just what we'd gone to Los Angeles to find.

But Dave can't come to the phone. He is dead, I explain. Dead since May. Too bad you didn't call sooner.

So sorry, says Walt. Silence. How about you, then? Me? Yes, you. No, I say. Not a chance.

Think about it, says Walt.

This is what I must have thought. I'm a thirty-eight-year-old widowed mother of four living in a partly finished log house on mortgaged land. Married for nineteen years, since I was nineteen, I have been the follower in this family, not the leader, following my husband from Chicago to Seattle to Missoula, even to Spain and Los Angeles. I pictured myself as the beta dog, a Sancho Panza to his Don Quixote—adapting, making do, working at whatever came to hand: editor, schoolteacher, social activist. Dave was the dream initiator, me the dream-come-true-maker. Never had I alone designed a house, created a film, written a script or book. My boys will tell you that like my mother, I was a kibitzer, occasionally nagging, full of ideas about the how and what of a project, whether or not it was mine.

Still, how could I refuse? If I refused, I'd be ashamed of myself and taunted by Dave's ghost. The job was my only offer, a better opportunity than any work I could find in Missoula. It was a way to pay off our debts, feed the kids, get medical insurance. And I needed to escape from the grief that mired me, like Blackfoot clay, in the ruts of defeat. On the ranch memories of Dave littered every bit of ground. There was no neutral corner. Nothing except some bell-bottom jeans and high-heeled boots to say, "This is me, mine, Annick's!"

Full of self-doubt but determined to try, I accepted

the job and went deeper into debt to buy a new car (my first) to signify the new me—a turquoise Subaru sedan with front-wheel drive (an innovation those days). I loved that spanky car, doomed as it was to be crashed first by teenage Steve, then by my Spokane Indian secretary, and again one blizzardy evening by a drunk in a jeep who spurted out of an alley, plowing into me and the twins on the main drag through Wallace, Idaho. The car met its demise from a snapped axle a couple of years later, with Eric at the wheel in a midnight black-ice skid along the Blackfoot.

In that gem-bright Subaru I'd fly across two mountain passes along the I-90 freeway, heading west two hundred miles from Missoula to Spokane each Monday dawn, then back to Montana Fridays after work. I had decided to commute because I was sure I wouldn't last long as a producer. If I'd be going home in a few months, I figured, there was no sense in heedlessly disrupting the lives of my fatherless boys.

Only three were in the nest that fall, for Eric had gone to work on a ranch in Colombia with one of his buddies, a trip perhaps inspired by the drug culture but opening new worlds at a crucial period of loss and awakening, a real adventure before going to college in Oregon. Steve seemed the most overtly wounded by his father's death, and I felt guilty about having thrown him into the pits of Hollywood High as a sophomore, so I acceded to his wish to finish high school in Missoula. The twins would attend second grade at the three-room school up the road in Potomac, with neighbor kids for friends and a trusted couple living at the ranch as housekeepers and baby-sitters.

It was not an ideal arrangement (looking back, I think I must have been a nutcake with too much energy and not enough motherly sense), but it was the best plan I could piece together. And it gave me freedom to try my emerging wings. From Monday through Friday I could work with no family distractions, learn what I had to learn, and face the challenge of producing a series of TV documentaries—a task that would become all-consuming.

In Spokane I rented a room in the house of another husbandless media mother and kept to myself. After years of tending a sick man and four boys, it was sweet relief to tend no one. In memory it is always November in Spokane, night streets slick with fallen leaves, silent houses inhabited by strangers, yet in that redbrick city of hills and a river, Catholics and vagabonds, I opened to the embrace of solitude as well as its tossing anxieties.

There were almost no Indian filmmakers in the mid-1970s, but after a long search I was able to hire an all-Indian crew headed by George Burdeau, a young director of Blackfeet descent. He brought in Larry Littlebird from the Santo Domingo Pueblo as a writer and codirector. Phil George, a Nez Perce poet and fancy dancer from the Chief Joseph clan, was on the creative team; and we enlisted two young interns from the Spokane Reservation. This was the height of the American Indian Movement, a time of awakening for many Native Americans, and also a heady moment for middle-class women like me, who were beginning to hold positions of authority and were absorbing a new, more consciousness-raising version of feminism from a younger generation.

Approaching forty but inexperienced, I was still romantic in my overeducated city-girl way about Indians and the West—a liberal wanna-be in charge of younger, ambitious, talented, and angry Indian men. At first we worked well together, all of us learning the ropes in the same wobbly boat. But as the crew grew more adept, confident, and unified, they resented my presence and position. I did not know what to do.

Eventually they threatened revolt. Some friends suggested I quit. Others urged me to work things out. Still others thought I should fire the lot and start over. Every night I wept in frustration. This was the toughest work I'd been confronted with, tougher than teaching high-school English or having twin babies. But I was no quitter, though perhaps stubborn to a fault. I finally did the sensible thing, appealing to the Indian grandmothers on my advisory board for guidance.

The grandmothers saved me. Tillie Nomee, in particular, saved our project and my ass. Tillie was a heavy, round-faced woman, her hair streaked with gray. She was a beadworking artist, a mother and grandmother many times over, and ill with diabetes. A respected Salish elder, Tillie had taken a leadership role in Spokane's urban Indian community, and it was my good luck that this wise and generous woman had befriended me.

Tillie told the guys to ease off. She told me to step back. She made us feel selfish and childish for wanting to feed our egos rather than advance our project for the sake of the seven tribes. The men and I made a kind of peace. Our show would go on cooperatively, on a more equal footing—me facilitating their creative work rather than trying to control it.

As my second year as a producer began, I'd become secure enough to rent a small, spare apartment near the public television station and to bring Alex and Andrew to Spokane for third grade. My littlest boys needed their mom, and I needed their warm, inquisitive, funny, and naughty bodies at my side. We had no furniture. Only mattresses on beige carpeting, a coffee table and a couple of chairs, some floor pillows, books, a few posters on the walls, a clock-radio, and a rented TV.

The coming year's road would be precipitous but at least now I had some actual miles on the vehicle. Our team had made a successful pilot film called, appropriately, *A Season of Grandmothers*, and we were working on the remaining programs in our Real People series. I would travel to New York to raise money, take the crew and grandmothers to the Washington Mall for a Smithsonian Native American exhibit, write proposals and budgets, go to public television conferences. In spite of obstacles real and imagined, I was becoming a producer.

The social life of a young-enough-to-be-attractive widowed mother of four is a strange item, indeed. There were the true friends who never failed me. And false friends who did not call or visit or invite me, afraid of the discomfort of being with my loss. And a few married male friends who guessed my need, felt my vulnerability, knew I was free to be pursued. But I did not want to betray my women friends and, perhaps in reaction, I found myself attracted and attractive to younger men. Sometimes much younger. They were transients in my life—good looking, sexy, and casual. It wasn't a case of wanting to relive my youth (I've never been nostalgic for those hormonic, intense, troublesome years), but young men were emotionally safe. Big love, or even some lesser attachment of the heart, was the last thing I was looking for. I knew these affairs would end soon, each mismatched partner tired of the other, with no words that could hold in the passionless light of day. Silent, distant mornings cured me of the young-lover habit. Celibacy would, at least, not be boring.

Then, about a year after my series was completed and I was home from Spokane, I became romantically entangled with the writer William Kittredge, who would become the companion and love of the second half of my life. If, when I began to keep company with him, anyone had prophesied such a turn of events, I would have shaken my head in disbelief, for Bill was so radically different than I that the distances between us seemed unbridgeable. My life was already unstable and chaotic. The commonsense solution for a widow woman like me would be to latch on to a steady guy who promised a prosperous future. Not some Rabelaisian Wild West autodidact with a genius for language.

Bill taught creative writing at the University of Montana. Dave and I had met him at English Department parties, and we both had been deeply moved by his story "Breaker of Horses," about an old cowboy's life and memories. But Bill ran with a hard-drinking bar crowd—CIA veterans and a race-car driver, carpenters and artists and gamblers, barmaids and divorcées and Realtors—whose do-anything will to wildness was beyond any partying we'd experienced.

Many writers drank a lot in those days. Whiskey seemed part of the writer's code, particularly writers drawn to Montana who felt they were kindred souls with William Faulkner, F. Scott Fitzgerald, Ernest Hemingway, or Dylan Thomas. The only serious drinkers I had known were writers: Nelson Algren in Chicago, Theodore Roethke in Seattle, and in Missoula our friend Richard Hugo.

I did not understand why drinking was part of the writer's mystique, for although I liked being high, I hated being drunk, hated thc loss of control and self, the dizziness and nausea and skewed perceptions, which may have been what others sought. In Bill's case, and in Hugo's, drinking was also part of the workingmen's cultures they came from—a way of socializing as common to them as dinner parties were to me. Which brings me to the politics of excess they both embodied, a Dionysian rebellion against a middle-class society that was puritanical in theory if not in reality, and too hypocritical and judgmental to encourage radical insights, much less radical change. Besides, we all know that alcohol is addictive.

Like Dick, Bill was a big man with big appetites, a writer and teacher with a thirst for knowledge and perfection. In background and personality, however, they were essentially different. Kittredge was no disenfranchised street kid from the blue-collar fringes of Seattle, but a man from the Oregon high desert, a son of big-time ranchers, an owner of land and cattle. Rough and shaggy maned, wide bellied, with his massive head and neck thrust forward when he walked (his nickname was the Buffalo), Bill wore his ostrich-skin cowboy boots like the westerner he was—a sometimes self-destructive, ironic, dispossessed, yet generous and loving teller of stories—a fit subject for the Hank Williams heartbroke country blues songs he admired above all others.

I wish I could say it was love at first sight—the black-haired Hungarian Jewish city girl and the blue-eyed son of pioneers—but it wasn't. Bill was the opposite of men who usually attracted me, the slender, athletic, fair-haired intellectuals like Dave. And his life was darker than any life I desired. I had seen his tender side only once, sitting on a couch in his house at a big literary party, when he told me of the breakup of his first marriage, and of the son and daughter he'd left behind for ten years. This was a story that touched me, but also puzzled me. In my family-centered universe such a thing seemed impossible.

I did not begin knowing Bill until I, myself, became familiar with bars, a practice that started with Hugo at the Milltown Bar and continued after Dave died, when I passed through Missoula on those Friday-night commutes from Spokane. Sometimes I arrived well past the twins' bedtime, so before heading up the Blackfoot, I'd stop for a drink and some fellowship at the Eastgate Liquor Store and Lounge. This was where Bill Kittredge held court under the velvet painting of an overripe nude (a gift from the Kittredge Collection) with the tall handsome young men who were his prize students in the 1970s: Ralph Beer, Neil MacMahon, Bryan DiSalvatore, Bill Finnegan, Fred Haefle.

I came to know the seductive pleasures of a bar during those rest stops, the warmth and laughter, the loosening and finally the shucking of other worlds and personas—a sense of freedom beguiling but false, for eventually even the hardest cases must go out those swinging doors and back into the daily realities we also inhabit.

Then one Fourth of July at the ranch when I was lonely and feeling sorry for myself (the boys were at their grandparents' cabin on Lake Michigan), Bill and our carpenter-writer friend Jon Jackson came to rescue me. We shared a jolly day of driving, playing, dining, and drinking on the shores of Flathead Lake, ending the long summer evening by dancing to the jukebox at the Trail's End Saloon in Missoula. Bill was warm and heavy with sexual energy as we swayed to a country ballad, and he began to sing in my ear, *"Here I go again / I hear those trumpets blow again . . ."* I felt my blood stir and rise. I fled home, alone, before it was too late.

Thus I was not surprised when, a couple of months later, Bill asked me out to dinner and a movie. I accepted, although I was scared, and rightfully so because the first thing that happened on our first date was a car wreck. Not Bill's fault. Some idiot ran a red light by the post office and plowed into Bill's yellow Toyota. But it was a harbinger of wrecks to come, for our flirtation would evolve into a full-blown affair that was torn with jealousies and recriminations and weeping and slamming of phones. But we stuck with it, because we discovered we needed each other. There was nothing boring in our mornings after.

I'll never forget a night early in our romance, when we bedded in Bill's tiny apartment. The ten-year-old twins were justifiably cranky, crammed into a sleeping bag on a black plastic Barcalounger on which an emphysemic, alcoholic old cowboy relative of Bill's had died years ago in Burns, Oregon. Bill and I lay in the next room under a reproduction of Hieronymus Bosch's triptych—*Hell, Paradise, and the Garden of Earthly Delights*—him chatty on speed, the radio tuned to a country station, me maudlin from too much Jim Beam, and we made love and confessed our secrets and wept and laughed the night through, sleeping in the hours before dawn and waking exhausted to a new day.

I'm confessing too much, the voracious self reveling in the delight of telling stories on oneself like an old granny going on and on until the children fall to sleep. Time and events merge in a swirl of memories. Years disappear. Kids grow like sprouting corn, grandkids are born, parents die, the forest is cut down, the barn roof caves in, there are film premieres and book tours and frightening days in hospitals, yet nothing seems changed in the bedrock self. No matter the wrinkles, white hair, love handles, and arthritic fingers, I believe I'm the same person I was back then, even though I'm not.

Memory functions outside of time, like fiction, movies, and dreams, and beyond the usual flashbacks to family and places and friends what comes most clearly to focus is work. Not the everlasting work of cleaning and cooking, or chopping wood or planting gardens or caring for children, dogs, horses, even cats, but the work of trying to make and sometimes, miraculously, making movies; and then the work of writing, which is far more satisfying than the work of being a writer.

Coming home from what had seemed exile in middle-brow Spokane, I joined with wildlife filmmaker Beth Ferris in filming a portrait of Dick Hugo. Dave and I had made a black-and-white movie about Dick ten years before, when the poet had been jilted, drinking, on the edge. Now he was married to Ripley Schemm, had adopted her family, found success, quit drinking, and counted himself happily settled for the first time in his life. We would film his new persona in color and edit in some of the black-and-white footage of the bad old days.

I too was feeling content for the first time in years. Making a movie in Montana with friends was a fine way to try my legs as an independent. The Hugo film was about love, not money, but Beth and I were cooking bigger plans. We formed a nonprofit corporation called Wilderness Women and, with a grant from the NEH and with associates Connie Poten and Claire Beckham, set out to produce a film series about women's lives in the historical West.

Revisionist historians at the time were beginning to recast the past in terms of family and daily life (a female version of history) instead of seeing only the conventional narrative about politics and power (the male version). But when we began our project in 1977, there were no movies or television shows that depicted the actual lives of women in the West. Western women were still portrayed as the "faithful pioneer wife" or the "shady lady with a heart of gold."

After three years of work, and with public funding, we produced and distributed a feature film called *Heartland* based on the life of Wyoming pioneer Elinore Stewart. It was written by Beth, I was the executive producer, and Bill was a consultant as well as cowriter. We filmed *Heartland* on the cold, snowy high plains of Montana and were gratified that it won many awards. But though the film has become a classic in the fields of women's studies and western film history, it did not make us rich. Not even famous.

Having produced a critically acclaimed feature, it seemed to make sense for me to move to Los Angeles or New York. This time it would not be a desperate mistake like the Hollywood foray with Dave, for now I had a track record, connections, and filmmaker friends. Producing *Heartland* had propelled me headlong into the independent film movement of the early 1980s, and I'd become a founding board member of both the Sundance Film Institute and the Independent Feature Project.

While being a Montana film person had a certain cachet (I could step off the plane at LAX or into a Manhattan meeting in deerskin boots and a brown-felt Stetson and immediately be noticed), such playacting is never enough in show biz, where everyone is costumed and wearing a mask. Trying to make it as a one-shot producer of a low-budget, low-grossing film while headquartering on a family homestead on Bear Creek Road in 1980 was Sisyphean. There was no film community in Missoula to offer advice and support, no way to schmooze with people in power, no crucial action—no action at all. And I was in a classic double bind. Though I dearly wanted to make movies, I did not want to leave home, or uproot my kids, or break up with Bill, who was tied to the university and wouldn't follow me in any event.

Besides, Bill and I were becoming a creative team, which was pleasing to a collaborative type like me, though he has always been a loner. Our most rewarding joint effort was working with writer friends such as Jim Welch, Mary Blew, Bill Lang, and Bill Bevis, to edit a huge anthology of Montana writing called *The Last Best Place*—a four year course in western literature. During those years Bill encouraged me to write. This was his strategy to keep me home, and he taught me some of the storytelling skills that had made him a sought-after professor of creative writing. I was an eager student, but the learning went both ways, for Bill was interested in screenwriting, which I could try to teach him.

We wrote several treatments together, but nothing came of most of our projects. Except the last one, which was the development of a film adaptation of Norman Maclean's novella *A River Runs Through It.* With Bill as writer and me as producer, we became partners with Norman, who held control of the story. Norman summered at Seeley Lake, about forty miles from my place, and during the five years we worked together the most important thing for me was getting to know the old man. I believe he was a great man: brilliant and funny and crotchety and generous and downright nasty.

The worst thing that happened (paradoxically, perhaps the best thing in the long run) was the intervention of Robert Redford. Norman was finally about to approve Bill's script when Redford's producer telephoned. "I have good news and bad news," he said. "The good news is Bob wants to direct *A River Runs Through It*. The bad news is Bob wants to direct *A River Runs Through It*." Redford courted MacClean and his family, then bought the film rights, acquired our rights, and made the film his way after Norman died.

Following Redford's takeover, I felt utterly disillusioned. I had reached my middle fifties and was looking at my hole card. A year or so later the system defeated me absolutely. The Cajun director Glen Pitre and New Orleans producer Eddie Kurtz had been working with me for several years on the development of a film for television about La Salle, the French explorer of the Mississippi River. After compromises and the seemingly inevitable loss of control and money, we were set to make it for Turner Television. Bill and I had packed his car and were ready to head South to New Orleans for preproduction when the telephone rang. Turner TV had nixed our deal along with thirty others because Gulf War coverage was eating up their company's cash and had priority.

I'm too old for this, I told Bill. Hell with them, he replied. We're going anyway.

New Orleans was a turning point. We would have a swell time in the French Quarter with Richard and

Kristina Ford, who lived a couple of blocks away on Bourbon Street, and it was in our rented shotgun house, people dancing outside in the streets, rain pouring, a world of blues, gumbo, and danger around every corner, that I decided to become a professional writer. Capital investment and collaborators were not needed, only time and a word processor. Best of all, there were editors who'd pay me to write the stories I was obsessed with, pay me to travel to Alaska, Budapest, Jackson Hole, the Chilean coast, pay me to observe whales and bears and butterflies—things I longed to do, would do for free if the chance arose. There would not be much dough in such writing, but enough to keep going, enough to reenergize a widow's life.

Then the windfall began. *A River Runs Through It* drove me out of the film business, but its profits have supported my writing habit as well as my living-in-Montana habit. And the five years of energy and time we invested in the project have, in this long run, enabled me and Bill to renovate my log house.

Transformation is nothing new for this place. The original structure was built a century ago by upriver pioneers who cut down virgin timbers, hewed them with adzes and double-bit axes; it was moved to our property in 1971 and rebuilt by Dave and me, the boys, and a few close friends; Dave died in its kitchen in 1974, but I stayed, and our kids grew up within its walls and have worked to keep those walls standing. The house is surely not mine but "ours"—Dave's and Steve's and Andrew's and Eric's and Alex's, and now Bill's too.

We started with the kitchen—the warm center. There would be a new tile floor, maple cabinets, gas convection oven, stainless-steel counters. I forgot to mention the incredible Culligan water filter that takes out even our well water's prodigious amounts of iron and minerals. Without that filter we'd be stuck with our old rust-streaked toilets, corroded pipes, water that stinks of sulfur and, when mixed with golden whiskey, turns the brew black.

So we tiled the downstairs bath. And replaced the cracked homemade windows with insulated new ones. And took out the woodstove in the basement, installing a propane furnace. This would be followed by a big deck looking out to the meadow, and deck furniture complete with umbrella. Then we literally raised the roof, tearing off the weathered cedar shakes that Dave and the boys and our carpenter pals and I had nailed up so carefully thirty years before. I wept to see that silvery moss-fringed roof come down, but the tin that replaced it is fireproof and carefree, providing safe cover for as long as I shall live.

Under the new roof we built a dormer study for Bill with picture windows looking down Bear Creek canyon to the north and south up the meadow to the woods and cliffs. Between the new study and my high-ceilinged bedroom/workroom, we replaced the dark, unfinished lavatory with a skylit Mexican-tiled sea-green bathroom where I soak every morning in verbena-scented bubbles in an oversized egg-white tub. And along the connecting hall we have actual closets for clothes and storage. A household comfort I never had before.

It seems strange to be crowing this shameless song of consumerism. Me, the back-to-earth hippie, celebrating shopping—scourge of the global economy. Me, the radical daughter and mother, shouting hallelujah for the delights of the material world! Maybe you'll forgive me. I'm sixty-five, a senior citizen, and can't help feeling it's time to enjoy all the pretty things before the dimness sets in.

I guess it's not so strange to crave comfort after living on the edges of poverty since I left home at nineteen. It makes me feel good that when my ninety-four-year-old mother comes to visit, I can offer her a fresh-painted room with a brand-new quilt; and I like to settle down with my granddaughter, Jessyca Rose, to watch a show on the digital Sony connected to a satellite—something my own kids could never do, because we had no television.

Comfort is nice, but it has little to do with serenity. For the last decade or so (call them the menopausal years, though God forbid not yet old age) I have been content to live in our white-chinked log house with dogs named Rasta and Little Red and Betty Boop and Bruno, and cats named Rascal and Ceca and Kevin. And with Bill, who shows up four or five nights a week. Bill is retired now, and mellowing like a vintage wine. He has reunited with his son, Brad, and daughter, Karen, and enjoys being grandfather to her sons—Zach, Riley, Max, and Leo who visit on the ranch some summers.

My grandchild, Jessyca Rose, is the great good surprise of these latter years. She is Alex's daughter, tall and dark-haired like he is, and lives with her mother, Robin, a few miles away in Potomac. Steve lives in Missoula these days and works in public relations. He fills our freezer with venison and smoked fish, and teaches Bill how to run his new computer. Eric is a prematurely gray Realtor and a world-class fly fisherman happily married to Becca—a woman I'd choose to be my daughter. They live a few hours away in Kalispell, with their golden retriever, Layla. Andrew writes poems and screenplays in Brooklyn, Alex writes fiction and screenplays in Los Angeles, and both are in Great Falls this severe winter codirecting a film they wrote, pursuing the lives their mom and dad set out on too late.

I hope I'm not boring you with this litany. Naming the names is important because family remains at the center of my life. But family is not the whole story. When everyone goes away, I am alone in the silence that is not silent. There are horses and cattle on the meadow, elk and bears in the woods, the hoots of great-horned owls at dusk, coyotes singing the moon. My heart beats to the pulse of life outside myself and I rest easy, I sleep.

Recently I went to my dentist to get a porcelain crown on one of the few remaining molars that grace my lower jaw. As age diminishes the physical attractions God gave me—teeth, skin, hair, legs—I become more concerned with protecting what's left. So I asked him why my lower front teeth are worn in a line that slopes from left to right. I was considering an implant, thinking uneven chewing had ground away those teeth.

No, he said. Your teeth are worn down from tooth gnashing.

"Have I been gnashing?" I asked, alarmed.

"Not recently," he said. "Those are old marks."

"I guess my teeth-gnashing days are over," I said, grinning.

His smile was whiter than mine, but quizzical, for what my dentist and I didn't say aloud is that the time of no-teeth-gnashing is fragile and much to be prized. Pain may come tomorrow, a root canal next week, and the probabilities of extraction increases each passing year.

Which brings me back to remodeling my house. Last spring the lease on Bill's flat was up for renewal. With his new study complete and almost furnished, he is spending more time at the ranch. Why don't I move up here, he suggested. A reasonable enough suggestion, one we've been considering a long time.

I think he saw the panic flash in my eyes. We've been together for twenty-three years, longer than either of us had ever been with a spouse; and though we've traveled the world, staying in hotels and condos for months, we have never lived together in one place called home. The prospect of Bill moving in seemed a change as big and permanent as marriage. I breathed harder. I hesitated. So soon? I asked. Both of us laughed. Bill renewed the lease on his apartment for another year.

Old age stalks in its tiresome details—the aching muscles and lost words and desire for security—but this summer, though fires burned all over our country, we pretended we were free as any blithe young creature, free to choose or not to choose. I know the shelter I have built and lived in with my extended family is as vulnerable as a bird's nest on the ground, and winter is coming. Soon choice will be no choice. But how can a woman live what she knows? The truth of a widow's life—the mundane truth of anyone's life—is take love any way it comes. One year at a time, one day, one breath.

PART 2

SINK OR SWIM

Indian summer. Montana. Rivers. You get the picture. Sky-blue waters gurgling with trout. Freckled cottonwood leaves like a yellow flotilla. Kingfishers, rose hips, purple asters. For me, eye-popping, soul-singing joy if that river is the Big Blackfoot a few miles upstream from my house.

By late afternoon on such a day, desk work is impossible. The river calls to me—a cathedral bell ringing—and I must go to it. The air is dry, hot, and blue, and the topsoil along the bluff is fluffy, like powder. As I walk my dogs up the Blackfoot from Whitaker Bridge, I turn my eyes from logged-off hills down to the cleft gorge where the snake of bright waters unwinds.

I am walking under ponderosa pines, trailing my fingers along reddish columns of picture-puzzle bark. The river is splattered with yellow light, but the water runs green. The Big Blackfoot is green as the scent of rain. Green as frogs and moss and evergreen trees. I crush pine needles under my feet, and their perfume mingles with the sharp odor of cottonwood sap and the trout scent of water over stones. Larches shower needles of gold on my head. These are the days of gold. I will dream of them during winter nights, the cold that may pounce, perhaps tomorrow, like a cougar from this self-same forest.

Betty Boop, my big-eyed shepherd dog, and Little Red, a bouncy young fellow part chow, part golden retriever, race ahead on the trail of a squirrel. Summer tourists are gone, only one jeep in the parking area. We head for a sandy beach that cuddles between arms of stone across from a great cracked tower of red rock. The rose-pink slabs rise like a Moorish castle above a tree-shaded pool.

I am sweating. The pool is inviting. I decide to take the last plunge of summer. But there on my beach, stretched out like a walrus on an aluminum lounge, is a huge naked man. He basks in rays of the sinking sun, and as my dogs sniff at his toes he raises a listless hand, turns his head toward me, waves.

I call my dogs and flee. Moving quickly through patches of poison ivy invasive as the nudist, I climb over a rocky headland, descend a fisherman's trail to a peninsula of silver-tossed willows. Betty and Red rush into the willows. There is a thrashing noise—a whitetail buck leaps from the brush and into the river. I watch him swim into the dappled current, head high, antlers gilded with sunlight. Buck. Naked. I laugh at the pure surprising coincidence. An osprey dives from the sky.

My dogs bark. They are barking at a raft with four people that is bobbing toward us from a bend in the river. Fishing lines extend on each side, like glowing cobwebs.

"How's it going?" I shout.

"Pretty good," says one fellow.

"Gorgeous day," says a girl.

"Anyone upstream?" I ask.

"No. I think we're the last."

Good, I think, for I know there's another swimming hole just ahead. My bank lies in shadows, but on the opposite shore a huge rock stack catches the light, its striated face patterned chartreuse and orange with lichens. At its feet is a sunny, graveled beach. This will be my perfect, private island.

When the raft is out of sight I strip off my shorts and T-shirt, leaving on my underpants and Tevas, and begin to ford the river. Ankle deep, the water is not too cold, but as I wade to my thighs it turns icy. I catch my breath. This is foolish. I should turn back, but I plunge in. I'm a strong swimmer. I have swum this river many days. But never in October. Never so alone. Never over sixty.

The current is stronger than it looks. I kick hard, turn my arms like windmills until I find myself in slower water. I crawl over slick stones—aqua, maroon, green—shivering toward shore. Although the sun still tips the great lichened rock and the pebbled beach holds the day's heat, I am cold, exposed, a half-naked crazy old woman clutching her arms around her flabby stomach, her breasts. My dogs have followed. They spray me with water, wag their tails.

Light is draining out of the canyon, so I don't linger. Going back I'll be smarter, I think. I will make my way upstream, then let the current carry me to the home shore, to my dry clothes, safety and warmth. But I've forgotten that the river is low. Slime-covered stones extend far into shallow water before the current takes over. I'm crawling crabwise again. And when I reach the deep water, it doesn't ride me homeward but curves downriver to the pool where I started my upstream journey.

I'm winded before I get halfway across. I kick and stroke, kick and stroke, going nowhere fast. My breath comes in gasps; the quickened beat of my heart pounds in my ears. Fear finds me. Not fear of drowning, but fear that my strength will fail and I will be vulnerable as a leaf, at the mercy of the river. Pull harder, I instruct my arms. My arms obey.

When I break through the current into a safe eddy, goose bumps cover my flesh; long white hair plasters my face; water drips from my eyes like tears. I stumble toward my clothes, the dogs frolicking and nosing at my weak knees. Even dressed and standing in a patch of sun, I tremble with the knowledge of mortality. I am older and weaker than the person I imagined I was, cold to the bone. Get moving, I tell myself. Get your blood running.

I start walking upriver on a piney ledge toward Belmont Creek. My head aches. My feet drag. It will take an hour before I feel halfway normal.

A mile ahead I see a jumble of sticks on a tall snag—the bald eagle aerie on Goose Flats. A white-headed giant screams at me as he glides from his perch. I want to scream back, "Hey, you, I'm still alive too!"

Here is what I learned from that Indian-summer Blackfoot plunge. Take no chances when you are dealing with a wild and precious river. Stay humble. The river is wider than it seems. Respect it. Death waits under slime-covered stones. Every living thing is mortal.

WILDCATS I HAVE NOT KNOWN

They are near to me. I have seen their tracks in the mud of logging roads and in the snow on Bear Creek's bottoms. The tracks are rounded. Four toed. With no claw marks, because the lion's claws are retracted when she walks. Her footprints are larger than those of the huge yellow dog who lives down the road, but unlike a dog she walks a straight line. Or nearly straight. The line of a stalker.

I have walked the trails the wildcats walk for nearly thirty years, yet I have never learned to think like a cat. My tracks are splayed, meandering. Some days I walk with my sons, or my companion, Bill, or friends, but usually I walk alone and always with dogs—the German shepherds I've loved like sisters: faithful Sylvie, black and tan, who came to the ranch when our kids were young; Rasta, alpha-bright and golden, who was dying of cancer but never faltered until her heart gave out by the cattail pond; her placid daughter Betty Boop, arthritic and gray muzzled at eleven years.

I have also walked with our black Lab, Shy Moon, who gave birth to forty-four mongrels; and Little Red Dog, smiling and dancing, who came out of the woods one snowy November to cheer us while Rasta sickened, then in the spring after she died ran back to the woods never to return; and now yellow-eyed Bruno, the bumptious chocolate Lab my granddaughter Jessy gave me to liven up my grandmother days. But in all my years of walking, I have never laid eyes on a mountain lion or lynx, not even the bobcats who also reside in these hills.

Maybe it's the dogs who scare them off—the old enmities—but I doubt if a mountain lion is so easily scared. The cat, as Kipling noted, walks by itself and, when it chooses, remains invisible. Perhaps that flash of amber up in the shadowy branches of a yellow pine is a wildcat's dark-tipped switchy tail. Or he's hiding under slabs of lichen-green rock on the cliff where I stand looking down at my meadow. I feel uneasy, vulnerable, stalked. I turn from the woods and head for home.

A scream in the night lifts the hair on my neck. It could be the scream of a fawn in the mountain lion's jaws, or it could be the lion's mating cry—life and death so connected I cannot distinguish one from the other. Fearful, I search for striped Rascal, the hunky neutered declawed tom the twins call dog-cat; and black-and-white Cica (pronounced *tse-tza,* meaning "cat" in Hungarian); and our inky, night hunter Kevin. This is dangerous country for housecats, and we have lost several to coyotes, owls, and lions. House cats are defenseless in the end against larger predators, but I prefer to let them roam, enjoying their true cat nature, than to confine them like prisoners in the house.

Looking into the cloud-black sky, no stars or moon or yellow lights of houses, I am surrounded by presences I love and cannot see, including the invisible cougars and ghost cats, and I am fully alive. The wildcats make me happy because as long as they inhabit my world, I know it is, at least in this way, wild.

The meadow on which my house sits is a circle of grass that has been domesticated like the horses who graze there and the summer cattle and me. We've had horses since our first spring on the ranch, beginning with the quarter horse Almaree, who birthed her red foal, Eustacia, in an April snowstorm. Then there was Steve's long-legged Hermano, and black Lupine, and three Shetland ponies who didn't last long. All our horses have long since died of old age or accident, and a horse-crazy girl from down the road runs her white mare, her feisty gelding, and a little Arabian over our meadow until the snow is so heavy they can't graze. I charge her almost nothing because I love to see those horses running and rolling, silhouetted on the ridgeline.

We have never owned cows (stupid, overbred creatures for the most part), but have leased the meadow for summer grazing to a neighbor rancher these thirty years. The cows stay as long as water runs in our ditches, which may be from May to July or May to September depending on the weather. They keep weeds down, help reseed the ground, remind me of the history of this place and the farmers who worked so hard to make the meadow productive. The income once paid my taxes. Now it doesn't come close. Cows for me are a habit or hobby, a matter of neighborliness, not of money.

Outside our circle of domesticity, however, lie miles of unpeopled country, which though roaded and scarred by logging is still the domain of wild animals. I sit in my hewn-log house and look out tall windows across open ground to a ragged, latticed line of firs, ponderosas, and western larches. The edge of meadow and forest is clear as a wall, but unlike a wall it is open. On my round oak table are binoculars through which I observe the beings that bridge the gap from trees to grass.

Whitetail deer emerge in early-morning light. Compared to the heavy-bodied cows they are delicate, pale, evanescent. Summer nights I catch their eyes gleaming red in my headlights. The deer have descended on my garden to munch the lettuce, browse my petunias. Lazy Betty and sleepy pup Bruno don't even bark. I honk my horn. The deer bound over the fence into the meadow, and all I can see is the white gleam of their upraised flags.

Often, I hear coyotes yapping. They have become bold because they know it is safe here. They play catch-me-if-you-can with my dogs, pounce on prairie dogs that burrow in rock piles. Elk appear at dusk in May, high-stepping out of the pines in search of small pink-centered flowers called spring beauties. The leader arches her neck like a camel, snout up to catch any scent of danger. Yearlings buck and chase. When night falls, some will bed on the meadow. I will see their impressions the next morning in the bent dew-laced grass.

In autumn, when tart wild apples hang red from the volunteer tree by my woodshed and the thorn apple brush is heavy with berries, we may be lucky at dawn or dusk to spot an elusive black bear, perhaps with cubs. We used to see so many bears we named a hollow in our woods Bear Cove. But they have been hunted and hounded and now I see only the steaming piles of berry-rich scat. I am content with that, satisfied the bears are still in the neighborhood.

My blood runs fast at the spectacle of wild creatures on my meadow or in the woods where I walk. I have been known to get teary at the sight of a traveling moose, a great horned owl, the bluebirds and kestrels that return with the runoff in March. Songs of spring peepers, the whirring whistle of snipe diving from the sky in their mating ritual, a drumbeat thrum of grouse, the red-tailed hawk who screeches as I pass his snag bring a kind of ecstasy, but I am still waiting for my mountain lion, my lynx and bobcats. Others have spotted them.

Once we had chickens. White leghorns arrived in a box as downy chicks, ordered from a catalog. Dave had sweet memories of waking to the rooster's alarm as a boy in rural Minnesota and ordered a dozen hens, a dozen roosters, thinking if one is good, twelve are better. Big mistake. The roosters formed a gang. They would follow the four-year-old twins, attacking from the rear. The older boys armed the twins with bricks, taught them to use a rusted frying pan to beat off the roosters. The roosters gang-banged the poor hens. We had to lock the free-ranging fowl in our henhouse to protect them. Eventually we rounded up most of the cocks. It was a bloody, stinking slaughter. Their meat was tough and gamy. Even when boiled for hours, my fricassee was inedible.

Winter that year was memorable for snow so deep we had to wear snowshoes to get from place to place. When the little boys walked the shoveled path from our cabin to the henhouse to feed the chickens, gather what eggs they could find, I saw only the red tassles of their stocking caps bobbing above the snow line. We decided to go to Seattle for a Christmas break. Evan, our neighbor down on Bear Creek, would feed the horses, the chickens, the dogs and Siamese cat.

We returned to find the hens gone. A bobcat had broken through the screened window of the henhouse, killed all the chickens, then settled in for a week to eat them one by one.

"I came up to feed 'em and there he was," said Evan. "Fat as a king. I took his picture. Figured the damage was done, so what the hell . . . may as well let him be."

The photo is blurry, but the marauder is clearly a bobcat. You can see his pointed, tufted ears, the spotted winter coat, slanted eyes, and white whiskers. All we could find of the prey were a few red combs and a dozen pairs of chicken feet.

Last fall, more than a quarter century later, Steve spotted two bobcats while hiking the ridge on Upper Bear Creek Road. He was with Betty.

"She treed him," he said. "She was all excited, barking

and jumping at the tree. It was a larch, most of its needles gone. That's how I could see they were cats. I looked up and there were two round faces looking down at me. They were much bigger than Rascal, the larger one maybe thirty pounds. Their fur blended in with the tree—a golden color. I could see the spots. They didn't make a sound or flinch. Stared me straight in the eye."

I was jealous. How many times have I passed that tree and never caught a glimpse of bobcats?

This spring I chat with my carpenter neighbor, Dan, who is driving his pickup through our place to cut firewood from slash piles left by Plum Creek loggers.

"Never seen so much timber coming off those hills," I say. "It's been log trucks all day, every day for months. I don't dare go walking up in the high country. Upsets me too much."

If I had high blood pressure, I'd be in danger of an apoplectic heart attack. As it is, avoiding destroyed habitat, I walk tight angry circles like a lion in a zoo. Which may sound silly to an urban person for whom 163 acres of grass and trees seems big as Central Park, but it's a fact of life for me.

"Took three days," says Dan, "for three sawyers to cut thirty-seven acres down to bare ground up toward Olson Peak. Just about made me cry. By the way, did you know you've got some mountain lions up there?"

He was pointing at the cliffs above my place, the only patch of tall timber left outside my own carefully guarded strip of forest. Mountain lions need large, isolated, game-rich hunting grounds to survive. I lease that state land and try to protect it as if it were mine.

"That's where I saw the tracks of a big cat," I say.

"Young ones," Dan continues. "Three of them."

Dan believes we are seeing more mountain lions because trophy hunters are killing off the prize males. Mature males are loners, often slaughtering kittens in their hunting territory, thus quelling competition and keeping the population in check. With the big daddies dead, unschooled youngsters proliferate.

"Loss of habitat," I counter, "is another explanation for more lion sightings and attacks."

As western Montana becomes the last resort for "white flight" families, movie stars, and the hardier variety of retirees, new housing developments are pushing into the wild woods and mountain slopes where lions dwell. What's a young lion to do when easy prey presents itself in his territory? A wandering poodle, for instance, or an overfed housecat; a plump toddler alone in the yard; a jogging woman who sets off his predator instinct, looking for all the world like a wounded doe.

"The problem's an explosion of people," I say, "not an explosion of wildcats."

I open the gate so Dan can pass through.

"Good thing we had an open winter. Should be a nice batch of fawns," he says, "and the elk'll be calving. No shortage there as far as I can see. That'll keep 'em busy."

Watching Dan's red pickup disappear into the Bear Creek watershed, I scan the meadow, which is dotted with mother cows and tender calves. There is no love in me for cows, but the calves, like all babies, are endearing. Moments like these I cannot help feeling protective of calves and fawns and spindly-legged newborn elk. Yet I am equally drawn to the mountain lions, who also must feed their young.

Nature is not confused with the mixed moral messages that perplex our self-reflective species. There is no pity in the world of eat-and-be-eaten. What I learn from the more or less wild place in which I live is respect for the beauty and logic of its interconnected parts.

There is new grass in April, berries in fall for deer and bear. Snipe build nests on the ground, and some nestlings feed owls and coyotes. To love the spotted fawn implies love for the wolf. Wildcats have killed my chickens, a housecat or two, perhaps even Little Red Dog, who loved to chase deer in the wildwoods. We take our chances. I will walk again this evening. Three lions live on the cliff. They are sleek, invisible, ready to pounce.

GLEANING WILD GARDENS

Cigarettes, ice cream, sex with strangers—maybe, for kicks, a roaring old-fashioned drunk—now, there's a fistful of guilty pleasures best given up these death-fearing days of the millennium. "Nails in your coffin," says Bill, as he inhales his mentholated True. I miss the sweetness of what I've given up. The smoke. The danger. "Who cares?" I reply. I'm fibbing. But huckleberries? Where is the sin in huckleberries?

In my part of the northern Rockies women (and quite a few men) have been picking the pungent wild blueberry for, say, ten thousand years—as long as the two species have existed in proximity. The Salish and Kootenai Indians who live in the Flathead valley on the west side of the Mission Mountains, just a couple of drainages from my Blackfoot valley home have gleaned their wild berry gardens for generations, reaping an important source of sugar and carbohydrates.

Gathering is the traditional work of women, and it includes the cultivation of a sustainable crop by selective picking and pruning. Old burns are a favorite place, for huckleberries, like prairies and old-growth forests, have a positive relationship with fire. The bushes need openings in the forest canopy to flower and bear fruit. Also, the nutrients returned to the soil after a burn fertilize the berries. If a patch isn't burned within seventy years, even the biggest bushes become sterile. In turn, the spreading rhizomes stabilize soil and hold off erosion.

Women of the berry-picking tribes know such things. They guard their patches in the way men guard fishing holes, and pass their knowledge to daughters and granddaughters—a practice not limited to Native Americans. My mother, in her nineties, still takes her grandchildren and great-grandchildren out into the blueberry patches that thrive in the loamy soil of the Michigan dunes, near the summer home we have frequented for over half a century. She talks as she picks. "I remember, when I was a girl," Mother might say, pulling her white, wide-brimmed hat down so sun will not freckle her soft skin, "there was a sour cherry tree in our orchard . . ." She tells stories of Transylvania, where she grew up. She remembers what is lost: her mother and her brother, and a gentler, slower, closer-to-earth way of life.

Grandmothers are the leaders of the cultures of gathering, and we have stories to tell. Some of them are love stories, for berry picking was a courtship ritual in many tribes. I like to imagine a young man and young woman (a version of me, with long black braids) wandering away from companions in summer twilight, a half moon rising, the undersides of clouds lit silver and apricot. There is a creek singing nearby where whitetails drink. The lovers hear a crashing in the underbrush. They freeze. It's a bear, aroused on the other side of the thicket. They drop their baskets, hold on to each other. Luckily, the bear stalks off in the opposite direction. But it's too late. The young couple has descended as one into the lush, sweet-scented, and sheltering brush. Their lips are blue rimmed and taste of berries.

Like sex, eating has remained pretty much the same over eons, and the consumption of berries is a cross-cultural ceremony. In the old days, as now, women stored their huckleberries in baskets or boxes. Women of the Northwest Coast mixed theirs with whale oil the way we mix ours with whipped cream. Plains-tribe ladies might have served them up with bison fat. Some batches were dried whole, or cooked to pulp and pressed into cakes, which were then dried for later use. Dried berries flavored stews, were mixed with bitterroot as a sweetener, or could be crushed into smoked meat to make pemmican—road food for travelers and warriors, or for automobile nomads like me.

I'd love to snack on pemmican as I drive toward Glacier Park with Steve. We will take the tour boat up Waterton Lake and hike three miles from Goat Haunt to the Kootenai Lakes. When we camped beneath the towering cirque wall a few years ago, a moose family, knee deep in marshy water, kept us up a good part of the night with their slurping. Next morning, up toward the pass, we found a bounteous huckleberry patch where we filled our plastic bags in half an hour. Walking out, we heard a noise in a great hemlock. Looking down at us was a fox-faced little mammal with dark fur, a rare and voracious fisher—the only one we have seen.

Huckleberrying can bring such unexpected gifts because while we are out in the woods looking for pea-sized fruit, alert to the presence of bears, our senses are jacked up. In the cities and towns where most people live, ten thousand daily cacophonies would drive us nuts if we practiced such alertness. One reason we go into the wild is to escape to a place where elements are basic and it is necessary to notice them. Old habits return, and like the animals we are, we tune in to actual weather. The terrain where we walk hides predators. We are as cautious as prey.

People like me retreat to secret places in the mountains whenever we can. Berrying gives us a good excuse. We load our old beater cars, our 4x4s and pickups with camping gear, kids, relatives, dogs, and head up to the hills, the trout streams, and the backcountry forests—what's left of them. We find a clearing in which to make camp amid ponderosa pines and clumps of white-flowering bear grass, and then, pails clanking, trail off into thickets and avalanche chutes, along streams, or up to the edges of clear-cuts and burns where we know the berries will be.

Family gatherings like these take place from Alaska to California, and from Oregon to Michigan—in the huge cool territories where more than a dozen species of wild blueberries grow. The most familiar to me is the thin-leaved variety that thrives in the damp, cedar- and fir-covered Cascade ranges of western Oregon and Washington; and the globular variety, which likes the drier, more acidic soils of ponderosa pine and larch groves in the northern Rockies. With greater moisture, Northwest berries grow larger than their Montana cousins, but the tough little Montana guys have a stronger, more concentrated flavor.

Several types of wild blueberries often grow in the same neighborhood. In Glacier National Park, for instance, six varieties intermingle: blue huckleberry, tall huckleberry, dwarf huckleberry, velvet-leaved huckleberry, dwarf bilberry, and grouse whortleberry. Only the blue and the tall are sought by grizzlies. And, as one woman researcher has discovered, the hucks become sweeter, more packed with nutrition as altitude rises. Which is a good thing for bears, who follow ripening crops from river valleys to high mountain ridges as autumn approaches.

Huckleberries ripen in late July or early August at their lowest elevations (about thirty-two hundred feet in western Montana) and keep on ripening into mid-September at elevations up to seven thousand feet. Hibernating bears must accumulate fat to survive in winter dens, to birth and suckle their young, and to stay alive through the snowy spring emergence. In my part of the northern Rockies most of their fat-producing calories come from huckleberries. The search for hucks usually keeps grizzlies and black bears safely in the backcountry, where they roam ever higher as winter calls, until the last berries are stripped.

Berrying is a favorite pastime in the mountain West, but now some alarmists are telling us to stop the picking, hold off on that pie, nix the huckleberry ice cream. We must boycott the commercially sold jams and the berry-filled chocolates in their Made-in-Montana wooden boxes. Why? Because huckleberries are getting scarcer in this region, and the bears who depend on them may starve, and thus the ecosystem is skewed (or screwed), and we're partway to blame.

The scarcity was palpable the summer of 1998 in western Montana, and again in 1999—although less severe. Bushes were barren, their bright green, ovate, minutely toothed leaves spotted yellow or turning a premature red. The few berries I found in my home patches were shriveled, dead white, inedible. This was true over much of Montana's huckleberry country, from the forested foothills of the Flathead and Swan valleys to the berry heavens of Glacier National Park.

The failures could have been caused by weather: premature budding brought on by February thaws; late freezes that nipped the tiny, bell-like flowers; a record thirty-two days over ninety degrees from July through September in 1998—climate changes that may themselves be caused by boundless human appetites for energy and more energy.

Bear biologist Charles Jonkel, a rosy-cheeked, white-bearded expert on bear behavior, believes that bumblebee pollinators flew in too late in 1998 to service the surviving early blooms, causing mass infertility. The same sexual miscues were likely to blame for the fruitlessness of serviceberry and chokecherry bushes, which abound in our valleys and hills, and usually serve as backup food for bears and other foragers.

"It's a five-hundred-year event," said Jonkel. He was sanguine about future crops. He pointed at Mount Sentinel, which rises behind the University of Montana. "Looks like we'll have a helluva season. See, the mountain's white with serviceberries in blossom." But a few days later, on Mother's Day, we had a twenty-degree night and woke to a two-inch crust of snow lying deadly white on Mount Sentinel's blossoms.

By harvest time, feeling the threat of starvation and with almost no backup food sources, a passel of bear mothers discarded their usual caution and came plundering into human territory. Instructing their hungry cubs, they raided orchards and gardens and garbage. Many sows were shot. Wildlife rehabilitation centers in Kalispell, Helena, and the Bitterroot valley took in nearly seventy orphans.

I saw for myself how desperate the situation had become when, in the summer of '98 (and again in '99), I went up into the larch and ponderosa and Douglas fir stands above my meadow to scout the berry patches where we gather each year and was astonished to find no berries, no bears. Not even bear scat. Climbing higher into Plum Creek's logged "industrial" forests, through clear-cuts and burns where the brush is thick, I still found no huckleberries worth taking. Poor bears, I thought.

They did not come down until late August. It was about three in the afternoon, sun shining on the sere grass. I'd been hanging sheets to dry (I love the ozone scent of sun-dried sheets) when I looked up to see a honey-and-amber panda-striped black bear sow and her two chocolate cubs walk out of the woods and across the meadow in plain view. They went directly to the thorn apple bush on the stone pile nearest our house and commenced the work of eating. I called Andrew, who was home visiting from Brooklyn. We whispered by the deck, binoculars in hand, old Betty on the lawn, dim sighted and oblivious.

The mama bear lifted her nose to the wind, seemed to look at us an instant, then returned to browsing. Her two cubs rose on their hind legs and pulled down branches with their paws, raking in the fruit through open jaws. We could hardly believe what we were seeing. In the thirty years my family has lived on this meadow we have spotted a number of bears in berry brush or raiding wild apple trees at dawn or dusk (our creek is, after all, named Bear Creek), but never at midday, out in the open. And during the last few years none of us has sighted any bears on the meadow—only their telltale piles of berry-rich scat.

The retreat of the bears has worried me. In the past I attributed it to hunting, for Montana has a black bear hunting season in spring. One ranger told me that too many mature bears have been taken, leaving untutored teenagers, not enough newborns. Add scarce food and disappearing habitat to hunting, and the sum is a major threat to survival. I wish I could protect all the bears, but I can only promise safety on my quarter section of land, which is a mere mini mart snack stop within the huge wild habitat they need to survive.

Through the rest of the summer our bear trio came again and again for their snacks, always in daylight, until the thorn apples were gone. I was glad to see they were still shy enough to keep away from the wild apple tree with its rosy fruit that grows a few yards from my house. I hope they found great thorn apple feasts down the brushy slopes above Bear Creek where grouse and wild turkeys hang out. I hope they made it through the winter. I hope to see them again this summer, but not starving at midday.

Whatever the causes of recent berry failures, we know that pickers have an impact on the huckleberry crop—especially in popular harvest areas where berries are accessible and lush. Some of the most prolific huckleberry gardens are no longer family secrets. Patches that nurtured rural, self-sufficient, earth-connected ways of life are being depleted by commercial pickers. The middlemen who hire them, and the businesses and markets that sell huckleberry products, are reaping profits from a resource that belongs to all Americans.

The gleaners scour brushy habitat near backcountry trails and roads in national parks and forests, often setting up camp. Many use homemade rakes or combs to strip berries from limbs, or beaters to shake them onto ground cloths, and some people break off loaded branches, hauling them to camp to be picked after dark. Although the picker may claim she is merely pruning, or wreaking no more havoc than an average bear, such methods wound or destroy the growth points at the tips of branches, leaving only the weakest shrubs to bear fruit the following year.

In addition to kids wanting to earn summer bucks and the usual itinerants, recent immigrants such as the Hmong people from Laos are becoming huckleberry entrepreneurs. After the Vietnam War, some Hmong tribes who had aided the CIA and American troops were transported to the United States for safety. One group chose to settle in Montana's Bitterroot valley and in Missoula. They've become meticulous and successful truck farmers, but like many old-time ranchers and farmers they seem to feel no need for stewardship in the wild. Unlike most other westerners, however, the Hmong are willing to pursue back-aching hand labor from dawn to dusk, bushwhacking into the wilderness to pick huckleberries for much-needed dollars. Because they are conscientious workers, leaving no bush unpicked, they become serious competitors of the black bears and grizzlies, grouse and songbirds that depend on the same huckleberry crops for sustenance.

Not until the early 1990s was the commercial picking of huckleberries even partially regulated on public lands. The wild crop seemed as endless as the spawning salmon that ran up every clear, gravelly stream. Now salmon are mostly gone from Oregon, Washington, Idaho, and Montana waters, but huckleberries are still here and ripe for protection. These days, if you want to pick hucks for profit in our national forests, you must get a permit from the ranger of the district you wish to enter. Day-use permits are four dollars (twenty dollars minimum) or eighty dollars per season, which is not prohibitive when pint bags are selling for five dollars or more in the Missoula Public Market. Wilderness and research areas, campgrounds, and developed recreational sites are off limits, and off-road vehicles and mechanisms that damage plants are prohibited, as is the severing of branches. But the consequence of disobeying such rules is only a ticket, carrying no standard fine or punishment.

Even if there were stiff fines, what agency can police the hundreds of thousands of acres of wildcountry available to pickers? Huckleberries are one natural resource that remains largely unstudied and unguarded—free for the taking from nature's lunch pail—and few opportunists can resist a free lunch, especially in an economy as poor as Montana's, where tourism is a growth industry offering a market for the million-dollar huckleberry business, which is value-added and regional—just what the economic doctors ordered.

The have-your-cake-and-eat-it answer would be to grow huckleberries as a crop on private ground. Unfortunately it is difficult, if not impossible, to successfully plant *Vaccinium membranaceum* (thin-leaf huckleberry), or *V. ovalifolium* (the oval-leaved species that grows alongside it in northwestern forests), or *V. globulare* (blue or globe huckleberry), the most common species east of the Cascades. Since habits of the wild berries' growth and production remain whimsical, unpredictable, and largely a mystery, domestication does not seem imminent. And a domesticated berry would not be as valuable. We might think of them as common, like supermarket blueberries. We would believe they had lost the taste of wildness so much to be desired.

Walk into the berry woods some bright early-August afternoon, with children or sisters or lovers or neighbors, and you will realize that we belong in this habitat. We are as natural browsing in the underbrush as rabbits and grouse and bedding does. Women especially seem at home here, humming as we stoop to pick the ripe huckleberries, tasting every so often because we can't resist the stomach's call.

Every summer for thirty years I have climbed into the rocky pine woods in the foothills above my meadow to go berrying. My neighbor took me the first time—an act of generosity, this sharing of territory—but my boys and I were no competition for her deft-fingered girls, who filled pail after pail while I scrambled after the wandering, eat-all-you-can-on-the-spot four-year-old twins.

Though I sometimes go alone, berrying is better done communally. Once long ago, when friends from Seattle were visiting, Bill came along. We had been drinking red wine, and he'd imbibed more than the rest of us. Bill is a burly man with a big head of thick, wavy hair, and when we found him under a prize bush, swiping berries into his open mouth, all of us burst into laughter, for if any person ever looked like a happy bear, it would be him.

Picking berries is not work but celebration. Even in a good year my family does not harvest many hucks, just three or four cottage-cheese cartons full, enough for a pie or two, some for cereal or pancakes or to sprinkle over ice cream, maybe a few jars of preserves. We might freeze a couple of pints for treats next Thanksgiving or Christmas—the scent of huckleberries in winter bringing the promise of summer to come.

For generations, there has been plenty of berry-rich country to make both bears *and* people fat. Loggers' wives in Libby, or ranchwomen in Drummond, or professors in Missoula made syrups and preserves from recipes passed down, perhaps, by pioneer great-grandmothers. Cooks in Florence or Arlee or Yaak baked all the huckleberry pies and berry-dotted pancakes they wanted, and hikers stuffed their pockets and rucksacks with impunity. In good years such small-scale picking still offers little threat to the berry supply and the bears who depend on it. The only solution in bad years is to leave *all* the hucks to the bears.

But the commercialization of huckleberries and demands from a growing market pose real dangers to a diminishing resource. Not long ago we experienced mushroom wars in the national forests—shootings and gangs protecting their special patches of morels or chanterelles. Soon, I'm afraid, we'll be having berry wars too, with competing gangs blasting away at each other on the public lands that are our nation's wild commons.

Boycotting commercial huckleberry products and the plastic bags of fresh-picked berries sold in markets is one way to send the message that we care about our wild gardens. Abstinence is another. And we could adopt stricter, better-enforced rules governing the gathering of wild plants, similar to the rules that govern hunting. No one argues that commercial hunters should be free to shoot all the elk they can find on public lands, then market them for profit. The world changes, and we must adjust.

These days Bill's hair is gray and mine is white, the boys are grown up, and we don't drink quite so much red wine. Still, when we get together this summer, we'll go berrying if it's a good year. There may be girlfriends, wives, a granddaughter, a great-grandmother. The dogs will be at our heels, and we will be singing or talking loudly to alert the bears, for if berries are ripe and abundant, bears will surely be nearby.

The air in early August will be dry and intoxicating: sun on pine needles, dust rising, earth musty under the low-spreading branches. Soon our fingers and mouths will be stained purple with the tart juice—mottled like the tongue of Little Red Dog, whose spirit haunts these woods. There will be animal scent in the air, perhaps a whiff of elk, or skunk, or the musk of a mother bear carried downwind. Bruno will be nose-to-the-ground on some trail. Old Betty will snap berries right off the branches. We'll shoo her away, but gently, this Huckleberry Hound who is like a maiden aunt.

Gathering wild food makes us inordinately happy because, I believe, it connects us to our deep past. To pick the wild blueberry is to take our place in a fast-diminishing ecology populated by plants and creatures that we did not create or domesticate—a connection impossible in Safeway. Food is a gift, no matter how it grows, but when we partake of wild food we feel the giving of it intensely. The berries fall ripe into our open palms.

You may think your own preservation has nothing to do with the preservation of huckleberries—or of black bears and grizzlies, grouse and wild turkeys and songbirds, or the small berry-eating mammals who forage in huckleberry habitat from the Pacific Coast to the Midwest. You may think you have nothing in common with Salish grandmothers and Libby housewives, or with me and my family on our meadow above Bear Creek.

Think again. If you are reading this, you are likely imagining something. Maybe you are remembering an old story your grandmother told. Or inventing a new story for your children. About bears. About picking berries. About hope that what is imagined and remembered may continue to exist, so that if we travel to where the wild things grow, they will be there for our gathering hands.

THREE WAYS OF WALKING THE BLACKFOOT

Start with the river as we dream it in myths and memory and movies. Green waters are rich with native cutts. The once-endangered bull trout swims dark and sinewy among the blue rocks and blood red stones. He lingers in shadows under lichen-painted cliffs, hovers in pools, patrols the beaches.

The season is spring. We walk among lavender pasqueflowers, magenta shooting stars, yellow glacier lilies called dogtooth violets. Tiny butterflies are blue clouds amid the flowers. There are hummingbirds and a kingfisher. A hatch of mayflies (or are they caddis?) floats to the sky like errant snowflakes. When the bald eagle takes off from a snag, we feel the wind of his wings.

Upstream from Whitaker Bridge, along the BLM's Blackfoot Corridor—ten miles recently saved from logging and developers—the watershed is not wild but nearly wild, land returned to the public, which is only proper. We turn our eyes from Plum Creek's logged-off hills, ignore the invasive knapweed, the leafy spurge and poison ivy. This is the good dream, after all.

The river is high and roily. Toe-sized salmonflies are wet with dew, sluggish on branches of white-blooming serviceberry. You can pluck the insects like winged fruit and throw them to the avid, slurping fish. Most likely the children are soaked (mine always were), hopping from rock to rock, tossing sticks to the chocolate-colored dog who is even more wet, waggy and barking.

Soon we reach the redrock cliffs and place a blanket on the sand. We will swim and picnic with the ones we love, and nap in the midday heat. Twilight washes the river in gold and silver, and we fish even if the fishing is bad. I wander upstream to sit on a rock and watch the water flow by.

The river sings to me of birth and renewal and solace. It is my hymn of innocent life and innocent death, as close as I get to prayer.

In my second dream I hear the rumble of experience. To witness it, travel east toward Ovando. We will make our way upstream from River Junction camp, where the North Fork joins the main Blackfoot, and bushwhack a mile or so through stumps in the canyon's remnant forest to where a huge hunk of cliffside slumped and sagged, slid, and finally tumbled into the river's bed. Blocking it. Damming it. Changing its course, at least in this short run.

The stretch of Blackfoot country around River Junction holds memories sweet and bitter for me. It is where, in a June season thirty years ago, my family found an abandoned hewn-log two-story house, a barn, and blacksmith shop. The buildings stood just downstream from the campground on a grass and sagebrush flat blue with wild iris. Dave decided we must have them. And so we bought the homestead for two hundred dollars and camped there for a month and tore the buildings down and stacked the logs on a friend's rig and hauled the whole shebang down the valley to Bear Creek, where we rebuilt it as our own. Now Dave is long dead, my sons grown, and the house, like me, has weathered, settled on new foundations, been remodeled for better or for worse.

If I could be sure the change in the river, the mudslide, the crash of clay and trees were also just an undercutting, a weakening, a transformation caused by weathers and waters and age, I would accept it as I accept the lined face that looks back at me from the mirror. But I am not sure.

The valley of the river that runs through it has been logged for generations. The great hewn timbers of my house were cut from a virgin forest a century ago. Anaconda logged the hills around and upstream from River Junction. Champion logged it more than once. And then came Plum Creek.

We used to camp at River Junction—a family reunion to celebrate the happy summer when we tore down the logs to make our house. My eldest son, Eric, who is a fine fly fisherman, and his wife Becca camped there as a newly married couple. He would wander upstream to fish those memory-laden waters. His last trip was a bad one.

"They logged it again," Eric said. "The place is a mess—it's ruined. I couldn't help it. I just broke down and cried."

He was referring to the land of the mudslide, logged and roaded for the fourth time since we have been witnesses. Having left the eroded earth nearly bereft of mature trees, Plum Creek sold it to a group of conservation-minded individuals who own riverfront property. We who prize the Blackfoot's watershed are lucky it wasn't sold to developers for subdivisions named River Runs Through It or Last Best Place.

Plum Creek's venture into the real-estate business is a new and frightening wrinkle. The company has reconstituted itself as a broker for its huge and far-flung land holdings—lands that once were given to the railroads (Burlington Northern was Plum Creek's parent). I guess waiting seventy years or more for forests to grow back in Montana's cold, slow-growing climate is not profitable enough for investors who want returns *right now!* Log as hard as you can, then sell the scabbed hills to the highest bidder seems to be the company's new policy.

Back above River Junction the land is in the river, which is running around it, trying to slice through. No one knows what will happen downstream if more slides like this occur on neighboring overlogged lands where the river curves and cuts into unstable banks that no longer have tree roots to hold the earth together. Will a load of mud and silt in high water destroy spawning beds and feeding grounds for endangered bull trout and native cutts? Will it be centuries before the Blackfoot heals itself?

With help from wetlands and fisheries experts, the new owners (many from out of state) are beginning the good battle of restoration—and, I'm told, the native fish are making a comeback. It is our good fortune these folks can afford to be so generous. Many cannot. The corporations, of course, should shoulder the greatest share of responsibilities for restoring the places that have made their shareholders rich. Future disasters could be prevented if corporations such as Plum Creek worked with those of us who live here to take better care of their (our) forests..

A river changes. That is the nature of rivers. The experience of rivers. We are part of that experience, and what we do affects processes of nature in a watershed. We call places like the Big Blackfoot River sacred because we cannot imagine life without their benevolent, healing presence. They hold our essential stories. The world does not have to be a devastated shore. As landowners and users, as a community, a state, a nation, we have the power to say *you must take better care of our common rivers, our common forests.* Failing that, we can simply say *no!*

Which brings us to the river's future. Will it be a nightmare? Or not?

The nightmare is mapped and charted, a vision of cyanide and gold. Come to the headwaters of the Blackfoot, up Highway 200 ten miles above Lincoln on the Lewis and Clark Trail, and we will walk the bad dream.

Here, just eight hundred yards from the Big Blackfoot, are timbered backcountry hills a mile wide and nearly a mile long. On the western borders of this patch of ground flow the Landers Fork and Copper Creek, where endangered bull trout spawn. A resident elk herd grazes the butte—that's where they are calving this spring. Sandhill cranes nest on this land, and wildcats prey on bounding deer.

Canyon Resources, a corporation based in Colorado, leased this ground from private owners and the state of Montana. Much of the ore it hoped to mine sits on state lands, which have been set aside to support public schools through income from grazing, mining, logging, or other developments. Their Seven-Up Pete Joint Venture also holds leases on two nearby sections even more wild, but decided to excavate its first mine on McDonald Meadows where we are walking. The company plans to level these rolling hills, to blast them with tons of ammonium nitrate, mining around the clock for up to twenty-five years.

The proposed pit would be a mile wide—about as big as the notorious Berkeley pit in Butte, where toxic waters still rise after more than a century of copper mining. The Canyon Resources pit would reach far below the water table, seven hundred feet under the adjoining Big Blackfoot River.

The company would pile the rock it calls "waste" along the river in a two-hundred-foot-high ridge. Imagine our landscape transformed with mounds of crushed rock high as a line of sixty-story skyscrapers. Heavy metals (sulfuric acid and nitrates) would likely leach into the Blackfoot, poisoning it as similar effluents from abandoned mines in the area have poisoned it before, leaving the upper reaches barren of trout and creating more EPA Superfund sites to accompany the ones already here.

But that's only half the nightmare. To separate the low-grade gold from this ore body, Canyon Resources would crush the rock, heap it in great stacks, and pour billions of gallons of cyanide-laced water over it. The water table would drop about thirteen hundred feet. Springs, wetlands, ponds, creeks in the area would become dry holes. For each ounce of gold extracted, Canyon Resources would need to process 245 tons of rock.

And to protect the nearby cattle and elk, insects and fish, children and farmers from cyanide, from arsenic leaching into wells and toxic metals leaking into the watershed, the company's plan calls for plastic liner sheets thin as a nickel, a technology that leaks poison every time it's used. Yes. That's it. Plastic and promises. The deadly words *trust me.*

The benefits of such a project (aside from profits going to out-of-state investors) would be fewer than four hundred short-term jobs earmarked mostly for experts from beyond the region. Trailer parks, land speculators, bars, casinos, and drug dealers would surely profit, along with a few local businesses and Montana Tech (formerly called the Butte School of Mines), which is the designated beneficiary of these state school lands. Those of us who live along the river would face increased and dangerous traffic, noise, sprawl, a poisoned river, ruined land, the demise of a rural way of life we love.

Huge industrial development at the headwaters of our green valley—or any other pristine watershed—threatens not only the environment but also traditional economies and communities based on agriculture, sustainable logging, recreation, and tourism. Boom and bust is a story Montana knows too well. We're still working on lots of unreclaimed disasters. Don't let's kid ourselves. We'll be digging into our pockets for taxes to reclaim the damage when the fat cats are gone to Canada or Barbados or bankrupt. It's the long range we're talking about. The reasons why we make this place our home instead of choosing Gary, Indiana.

The people of Montana are not foolish enough to buy a dream this bad. During the elections of 1998 they overwhelmingly approved an initiative that outlaws the use of cyanide in any new mining project, thus stopping the McDonald mine and other proposed heap-leach mines *for the time being.*

I say *time being* because Canyon Resources has not given up. It has initiated a lawsuit claiming that the anticyanide initiative, which has become state law, is an unconstitutional "takings" of the company's investments and future profits. It is asking for six hundred million dollars in damages or failing that, they want the ban overturned. This lawsuit is a blatant challenge to the democratic process, an "in-your-face" move that, if it succeeds, would validate any citizen's most cynical notions about politics. Courts, however, do not lightly overturn the peoples' will. So the mining industry, and politicians who are in its pocket, are talking about a counterinitiative to reinstate cyanide. And our avidly pro-industry Republican governor, along with a majority of state legislators, have recently rescinded environmental restrictions on energy development, logging, and mining. Soon out-of-state political action groups are likely to stage a media campaign like the one that defeated our Clean Water Initiative in 1996, where nine dollars were spent for each negative vote.

Until recently the price of gold was low, providing little incentive for Canyon Resources. The company closed its office in Lincoln, and dismissed its on-site staff. It is in trouble with the state for failing to clean up another mine, and its financial backer has sued for return of its money. This is truly one bad outfit.

It feels good to think we have won the struggle to save the Blackfoot, but we can't be smug. Gold prices fluctuate. Mining people claim that 9.9 million ounces of gold and 30 million ounces of silver lie under McDonald Meadows and its adjacent wildlands. If Canyon Resources fails, some other company is likely to try in the future. Miners are not about to give up or forget about a potential fortune. As long as precious metals sit under public lands, and as long as people are willing to pay for the shiny stuff, the danger will continue.

I am often asked if I'm against mining per se. Of course not. I drive a car, cook on a stove, use a word processor. I even wear jewelry (although I'm boycotting gold). I'm just against mining in places whose natural attributes are more valuable than gold. And I am not alone in this. About 60 percent of Montanans voted for the anticyanide initiative. They are not against jobs or school funding but, like me, are *for* a healthy environment, for ranching and fishing and life-giving rivers that belong to all Americans.

The people of this state hold the power in what is still a democratic process. We must fight for our last best valleys and to keep our clean waters. Come, take the next step. Rafters, ranchers, loggers, parents, tourists, outfitters, New Yorkers, New Agers, students, fisherpersons, even Californians—let us act now to support our great and common river. The future.

THE BEST LAST PLACE

OLD CEMETERY IN THE YAAK

The hill isn't much, just a slope in the tall green of the Yaak valley, a cleared patch marked by sentinel larch maybe two hundred years old, maybe twice two hundred. The trees are arrow straight, tempered by fire. This June afternoon they are clad in the uncanny chartreuse colors of spring and their newly emerged needles are as soft as a feathered boa. Looking skyward through the canopy of old growth, I watch branches rise in gossamer light like souls seeking release, which is an easy notion, for the roots of these trees are nourished by the dead.

"Hug a tree," says young Scott, a recent Yaak immigrant who is working to keep the roadless forests wild. My arms encircle less than half the girth of an ancient giant. My cheek rests on its bark. I breathe the perfume of sap and decaying needles. Looking up through green lace to a sky fractured into bits of blue, bits of pearl, I laugh because I would rather be walking this teeming earth than aspiring to heaven.

I am a tree hugger, not a tree chopper like most of the folks who lie in the Boyd Hill Cemetery, people who fled into the Yaak when the going was rough in more civilized parts, or when going was the only way. The folks buried here could not have wandered into the Yaak by mistake because it is a nearly impenetrable corner on the Montana/Idaho/Canadian border where the maritime rain forest meets the Rockies, and it is not on the way to anywhere else.

You must arrive in the Yaak on purpose, and not without difficulty. Those who lie in its final resting place came to hunt deer, elk, and grizzlies, or to trap wolves or beavers, or to fish for red-throated cutts and bull trout. Some came to homestead—scratching out gardens and reaping native grasses in the meadows along the river—but most of the people whose bones lie in the forest's ground came to cut down its trees. What a dreamland it must have been for those original loggers—a seeming infinity of trees. The forest called. They cut a hunk of it down. Now they *are* the forest.

Within the fenced enclosure, hand-chiseled names are hacked into blocks of stone. Families go back to the 1880s. There are veterans of World War I and World War II all in a row. There is Froy, and Wife of Froy—no first familiar name to mark her passing. Some metal grave markers once held pictures, but the oval frames are empty. Even the nameless are not forgotten. I see flowers fresh from Memorial Day and plastic bouquets: electric blue, shocking pink.

One plot is set apart by a white picket fence. Perhaps to keep predators away from the babies. Minuscule graves take only a small patch of earth. So many little ones dead in one family. I stand above the markers, amazed. This is not child death from the influenza of 1923, which is a common sight in old burial grounds, but death come in 1985, '87, '89. You could weep for that mother. Weep for her brood. Weep for your children alive in their troubles—weep for your own good luck.

The *rat-a-tat-tat* drumming of a pileated woodpecker breaks the silence like a chain saw. *Time to go. Time to go.* We pile into Scott's 4x4 and head for the crossroads called town. *Yaak,* according to a Kutenai tribal story, means "arrow."

"The Kootenai River is curved around our valley like a bow," says Scott's wife, Sherrie, who makes her living massaging tourists. "The Yaak runs straight down the middle."

I like to imagine life is a river. If life is a river that curves like a bow, stories are its arrows, and we are the slingers. When the stories of the living are joined with the stories of the dead, what you've got is a culture, a place that holds meaning, a home.

In Boyd Hill Cemetery I was touched by a world larger than the sum of meadows and mountains, grizzlies and trout, old-growth cedar, waterfalls, and white-starred bunchberry dogwood. I cannot know the stories a trout knows, although I am sure each trout has a story, as do the migrating eagles, the wolves come down from Canada. I will never think like the mountain lion or moose, but I can know the stories of humans.

There are people in this valley descended from the pioneers whose remains nourish larches on Boyd Hill. There are communal hippies who settled here in the 1970s, also foresters, teachers, nurses, radical freedom seekers from the far right and the left. I chat with a tattooed trapper and a white-permed retiree, a survivalist gun toter and college-bound teens.

We are sitting in the Dirty Shame bar, which holds more stories than the graveyard. I recall one gravestone inscribed PISS FIR JIM. The red enameled image of a chain saw was etched into the marble slab. Jim died not long ago. He must surely have been a regular at this watering hole. I try to imagine him sitting at the bar in high-heeled logging boots made in Spokane, black pants rolled at the cuffs and held up by suspenders. I ponder his message. Was Jim a logger who pissed on firs? Or did the firs piss on him?

Such speculation is foolishness. I know as well as the next person that "piss fir" is a kind of tree—a poor kind of tree—the kind a hard-up logger might cut after the grandfather ponderosas and cedars and larches and Doug firs have become roofs and decks and tongue-and-groove flooring.

Downing my beer, I order another. When twilight fades, I think I'll go out beyond the small perimeter of houses into the dark woods. I will follow a dead logger's implicit advice about how to grieve in the Yaak. My plan is to squat beneath an old fir, hold up my glass for Jim, pull down my jeans, and piss.

When I'm on my journey, don't you weep after me ... the words come back as I weave unsteadily into the woods—lines from a folk song we sang over my husband's grave more than twenty-five years ago. We were burying him in the Catholic cemetery next to the Northside softball fields in Missoula, where Dave had loved to play ball. The Dirty Shame and a few nearby houses cast a yellow glow along the two-lane. I turn my back to the light. In the other direction lies wildness. I step toward the dark.

When it's my time to go, please folks, don't weep after me. Just bury my body in a pine box set near to a big old tree. I'd like the tree to be a western larch in a remnant old-growth forest where I have lived and loved. My best last place will be a patch of earth preserved like Boyd Hill Cemetery, where coyotes sing and elk bugle and bears burrow, and where, at last, I can become worm food, joined in every atom of nonbeing to the energy that runs from earth to sky and back again.

PART 3

THE WHALES ARE SINGING

It is November in Sitka, on Baranof Island, along the narrow southeastern panhandle of Alaska. November sweeps into the fjords and the thousand humped islands and lies down among the moss-covered deadfalls and walks in the muskeg that springs under my boots like a succulent mattress. November rains on hemlock and spruce. It patters on trollers and tour boats and Tlingit totem poles and onto the patch of water where a whale is singing.

The humpback is suspended under thirty feet of dense green water. His fifteen-foot-long pectoral fins are opened like arms waving, like wings of an underwater angel. This is why his genus is named *Megaptera*—great winged. The fins are white, outlined in black, as are his symmetrical flukes. I am assuming this adult, who is as big as a bus (thirty to forty feet long and weighing almost forty tons), is male because it is singing. So far, only males have been identified as song singers.

Music is a connection that draws me toward humpbacks as it has drawn me to the males in my life. When I was a child, my father conducted Verdi's *Requiem* in our living room, waving his arms and singing along with the hi-fi. I waved my arms copying him. My four sons turn on the stereo each time they enter our house. They write to music, eat to music, and, I suppose, make love to music. When I was eighteen Dave courted me by taking me to hear *Don Giovanni*. Years after Dave died, Bill drove me to Alabama on a Hank Williams pilgrimage.

"Music belonged to sound," wrote Patricia Hampl about Antonín Dvořák, "the endless voiced revelation of God's nature."

In the summer of 1893, on vacation, Dvořák tracked scarlet tanagers through the woods and meadows of Spillville, a tiny town in northeastern Iowa populated by immigrant Czechs. He was trying to precisely notate the vivid song of the vivid bird. His research became transmuted and transformed as a major theme in the Scherzo of his Opus 96 string quartet.

By *song,* I mean an ordered sequence, a melody, a rhythmic pattern, a created thing. Wind in pines, thunder, waves breaking on a pebbled beach are not songs, no matter how beautiful their sounds may be, or how evocative. But like Dvořák's tanager, the canyon wren sings an ascending melody so distinctive that I remember the first time I heard it. Its notes transport me to the red-walled desert, the turbid roisterous river, the cathedral of the Grand Canyon.

The song of the humpback is equally brilliant, although far more complex. In time I will respond to it intellectually, but now, riding the slate-colored sea, sound waves bounce off our small vessel and I tremble with joy. My breath catches in fear. The only way I can explain this emotional connection is to call it primeval. According to whale expert Roger Payne, I may not be far off the mark, for at some far-past, evolutionary blood level, he believes we share a common heritage with both birds and whales.

I am sitting in the bow of Jan Straley's fiberglass whale-research boat, fast enough with its seventy-five-horse outboard to track humpbacks. Jan is in her mid-forties, a blue-eyed pioneer researcher, who has been following Sitka's whales for twenty years. She captures them with her three-hundred-millimeter lens, photographing the high-rising black and white flukes as the whales prepare to dive. The fluke markings are as distinctive as fingerprints. "I've tracked some whales since they were calves," she explains. "They come back year after year. Now they're grandmothers."

Jan has recorded humpbacks in Sitka singing snatches of song in summer, and others singing full songs in autumn and winter—songs that researchers once believed were sung only by courting males, and only in winter at the herd's warm-water breeding grounds in Hawaii or Mexico. As winter approaches, Jan has also observed butting, slapping, and other aggressive behaviors associated with courtship. "Their hormones are kicking in," she says. "It's the testosterone. Sexual changes don't happen in an instant. Not in a body big as a whale's."

Humpbacks singing in winter is what drew me to Sitka in the rain-soaked month of November. By *singing,* I don't mean vocalizations such as the shrill trumpet calls of the hunt; or the chirping chatter between mothers and calves; or warning cries; or clicks and hoots that fill whale-inhabited seas with sonic talk.

The whale we hear near our boat is singing a formal song. If I were underwater, sonic waves would vibrate through me, shiver me, resonate within me. I might hear whale songs from five miles away (researchers estimate that whales can detect songs sung more than 100 miles away and, like elephants, can pick up low frequencies inaudible to humans.

Jan's hydrophone is recording low moanings, snores, and rumbles, as well as modulated wavers—*oos, ees, whoos, foos.* There are abrupt changes from wavers to chirping sounds and tweets.Then a series of higher-frequency *yups, mups,* and *ups.*

This song lasts about fifteen minutes (some can run for half an hour), and then the whale repeats it. He sings as before, stopping at the same phrase to come up for air. He may sing more than one song. He may sing nonstop for hours.

Sonograms of humpback songs depict the shapes of sounds as well as pitch and duration.To me, they are as whimsical as a Klee design or deeply modulated as a Rothko oil. One sonogram sheet that I examined depicted lower-register phrases that resembled ink blots and were continuous as a fog bank. Other patterns, rising from low to high, repeated in diminishing sequences, looked like sketches of Christmas trees, or thumbprints, or quarter notes on a musical score.

A humpback's song consists of themes, phrases, variations, and transitions, often repeated in sequence. It is more complex than any other nonhuman vocalization we are aware of. Some listeners believe that phrases in a humpback's song rhyme, and that the rhymes change as the songs evolve. They believe rhyming may be memory aids, like repeated phrases in oral-tradition stories.

I love the idea of whale epics, whale sonnets. I wonder what, if anything, the humpbacks are remembering. What creation story has evolved over twelve million years since they became distinct from other rorquals? Has that story changed since 1996, when they faced extinction? The population, down to fewer than 10,000 then, is now estimated at 150,000 worldwide.

Walt Whitman said a poet's work is "to indicate the path between reality and men's souls." The humpback's song is a path to a reality outside our experience, emanating from a creature who some of us believe must have a soul to create works of such complex beauty.

Jan Straley is fascinated by whale behavior to the point of obsession. In 1979 she and her husband, John—a private investigator and detective novelist—summered in a Forest Service shack on Admiralty Island east of Sitra. Jan, a field biologist was planning to study geese, but no geese appeared. Instead close to one hundred humpbacks took up residence within a two-mile radius. "It was like a sprinkler system had cut loose," she says.

Observing humpbacks became Jan's lifework. Her son, Finn, was born while she was studying humpbacks at Glacier Bay. He spent his babyhood in a boat. "Maybe that's why he goes to sleep as soon as we get under way," she confides. "He thinks he's being rocked." Maybe, whale songs are his lullabies.

Our submerged singer has ended his song, and Jan and I are scanning the sea for more whales. Flocks of mountains, glacier white, rise around us like startled swans. Westward, on the island named Kruzof, Mount Edgecombe's cone is skirted in snow. I am reminded of Mount Fuji gone wild with eagles and bears and the tiny Sitka deer. Near the island's shore we spot a spattering of whale spouts, which appear and disappear like movable fountains.

If humpbacks were as predictable as monarch butterflies they'd be traveling south now, toward the warmer waters of Hawaii or Mexico, to sing, to breed, to calve. Although some herds in the North Pacific group have already embarked, the two to three hundred who feed around Sitka from September through January are not going anywhere. They stay for the herring.

In Sitka Sound in late November and December, great schools of herring seek protection in the fjords of Baranof Island. Humpbacks can stick around to gorge themselves as late as January and still make it to Hawaii in time to breed and calve. Jan and her compatriot Chris Gabriele (who studies humpbacks in Glacier Bay and Maui) tracked one guy more than three thousand miles in thirty-nine days—including side trips and rest stops.

I ask Jan if the song we recorded is the same one our invisible male will sing in Maui, and how it differs from the songs of other males. "It's probably the same," she says. Sonogram tracings have shown that each of the three geographical groups in the North Pacific sings basically the same song (with styles as distinct as a Sinatra version is from a cover by Elvis). And every summer the song changes. Not totally, but in its parts.

"Like Bach variations?" I ask. Jan shakes her head. She's a scientist. She does not speak in anthropomorphic similes.

"Is there one composer whale?" I continue. Jan shrugs. No one knows which individual or group initiates changes. We do know that almost simultaneously all the males in a herd will incorporate the same new themes and junk the same old ones. And we know that these changes will travel to the breeding grounds of other herds, until all the males in a geographical group will be singing the same variations. "But we don't know how," says Jan.

Whale research is a new science. There's a lot we don't know about the humpback's song—beginning with the act of singing—for although their voices have been picked up in space by *Voyager* and *Galaxy,* whales do not have vocal cords. Biologists surmise that their vocalizations are produced by vibrating sinuses, the bony caverns of a whale's head serving as sounding boards, but no one has proved how the process works.

Some researchers believe that males sing to attract

females, the songs a courtship display like the great horns of ungulates (evolutionary ancestors of whales). But singers tend to attract males rather than females. Which prompts other biologists to speculate that the song is a strategy to divert the male who escorts a cow and a calf so other males can get at her—a behavior that seems selfless, like a humpback's cooperative hunting techniques. My favorite theory is that the singing is a test of dominance, the best performers winning the gene-pool survival race.

Transmission is even more of a mystery than composition. Some researchers think a song's seasonal changes are transmitted by memes (neural codes created in the brain when an individual learns a behavior), a concept we understand as cultural tradition or, more simply, as ideas. A few nomadic humpbacks would carry the changes from herd to herd, like bards. And their audiences would, accept and mimic the new variations.

I prefer the "bard" concept to the one that explains transmission by a theory of genetic programming in which humpbacks are hard-wired to make progressive changes in their songs. This theory would solve the problem of similar song changes at nearly the same time over widely dispersed areas, but it is questionable, since no behaviors of this sort in other animals have proved to be genetic. Another objection is that since groups in the North Atlantic, the Southern Hemisphere, and the North Pacific sing distinctly different songs, it's unlikely they are carrying out specieswide genetic instructions.

Yet another theory posits that "rules" directing how to change song structures are taught to juveniles by their elders. In this way a scattered population could share a cultural tradition passed from generation to generation. Thus, cultural evolution and rule teaching, as well as some degree of genetic programming, could work at the same time, as they do for us.

I throw up my hands. Theorizing drives me crazy. I ask Jan what she likes best about studying whales.

"I love being able to follow a whale's behavioral history. I'm interested in the communications between mothers and their young," says Jan.

Me too. Writers and scientists share an interest in blood connections. In sex. In songs of life and death.

Sleet drives against a seaside-view window. Jan and I have ended our day in a restaurant and are eating warm seafood salads, drinking zinfandel, speculating about why so many young women have become field biologists and are studying whales.

"I think we identify with them," she says. "They live the same life span, and mothers take care of their young like we do. Besides, the females are larger than males. More dominant." We laugh. "Being with whales makes me feel very, very small," she adds. "They let me know where humans stand in this huge world."

Jan's husband, John, has a different take. "I think Jan's attracted to whales because they must be wild forever," he says. "She loves the fact they can't be domesticated."

I agree that whales are wildness embodied, and I too am in love with the wild, but other creatures are just as wild (if not so huge). Singing is the trait that draws me to humpbacks. The song of our singer echoes in my being. I'm as enchanted by his aria as I was when I heard the canyon wren's sonsinger echoes in my being. I'm as enchanted by his aria as I was when I heard the canyon wren's song, or a Bach cantata, or my father whistling a Dvořák scherzo. It pleases me to think that the tiny wren, the big-as-a-bus whale, and me—a five-foot-six-inch upright walker—may have a vertebrate connection that manifests itself in music.

Whales, like birds, left our earthbound nature millions of years ago, and their songs come from a consciousness beyond our comprehension, an otherness we call wild for lack of a more specific concept. Why, then, am I moved to tears by the humpback's song? Music is the common ground, I believe, the cord that binds humans to birds and whales. We three alien but familiar species sing the world's connections. Each in our separate ways, we sing the old songs of air, earth, and oceans.

SUMMERING GROUND

Tonight the air is teal blue, so luminous it seems a Venetian heaven concocted by Tintoretto. Above the mirroring lake, white peaks and serrated ridges are lit by alpenglow. This is no dream. The wooden stoop is splintery. My feet rest on grass. Behind the screen door, Bill calls me to bed, but I will stay until the hotel's yellow lights blink out. I will wait for stars to emerge, watch trout rise, hear the flutter of ducks in the winging silence.

It is the last night of June and I am in the heart of Glaier National Park at the old rambling chalet called Many Glaciers. Beyond my heavenly mountain-ringed valley, in alpine backcountry camps, in roadside autoparks sheltered by old-growth cedars, in motels and historic hotels, travelers like me, some two million each year, are also finding the solace they have come to expect on this eco-island of glaciated rock.

Glacier Park's 1,583 square miles are the northernmost reach of Montana's Rockies, spreading an additional 203 square miles into Alberta's Waterton Park to form an international peace park—a World Heritage Site. There are forty-eight glaciers in the park, all receding. Within fifty years, given the accelerated warming of the atmosphere, there will likely be none. But it's the great glacier swarms of the ice ages that give the park its name. They pulsed forward and retreated in recurring waves during the Pleistocene period, from two million to eleven thousand years ago.

The hanging valleys, cirques, and amphitheaters that mark the park's high country were sculpted by ice—as were the jagged knife-edge ridges called arêtes and the U-shaped watersheds descending from the Continental Divide. Thousands of tons of compressed moving ice ground rock into rubble, plucked house-sized boulders from bedrock, deposited mounds of moraine. Glaciers scooped out troughs, then the melt filled them, creating myriad forms of captured water. There are fjordlke lakes such as McDonald, Saint Mary, and Waterton, which run between seven and ten miles in length and reach depths of more than four hundred feet. Smaller lakes such as Bowman and Ellen Wilson call fishermen to try their luck catching westslope cutthroat trout. Chains of lakes are called paternosters because they are strung like rosary beads, and the parklands are dotted with hundreds of tarns.

Many of the ponds, streams, rivers, and lakes are colored aquamarine, apple green, robin's-egg blue, milky jade. This too is the work of glaciers, for suspended within the water is a fine silt called rock flour, which, when illuminated by sun, radiates colors not otherwise seen in natural waters. Frame these glacial effusions with wildflowers. Populate their banks with moose, elk, wolves, bighorn sheep, and grizzlies. Add Canada honkers—also mallards, eagles, mountain bluebirds—and what you have is a vision of paradise.

John Bird Grinnell, an early-twentieth-century author, editor, and conservationist whose enthusiasm helped bring the park into existence, called Glacier the Crown of the Continent. Lewis and Clark, on seeing the Rocky Mountain Front rise abruptly from the Great Plains, referred to the jumbled peaks as the land of shining mountains. To the Blackfeet tribes who controlled the buffalo-rich grasslands that slope east from the mountains, the Backbone of the World was a sacred terrain where men on vision quests might climb to the top of Chief Mountain. The highlands were also a place of myths. One Blackfeet story tells us that Old Man, Napi, who created all things, ascended to the sky from Going-to-the-Sun Mountain when his earthly work was done.

John Muir rhapsodized about Glacier in his 1901 book, *Our National Parks.* "Give a month at least to this precious preserve," he instructed. "The time will not be taken from the sum of your life. . . . Nevermore will time seem short or long, and cares will never again fall heavily on you, but gently and kindly as gifts from heaven."

Lucky for me, I live in Montana about a three-hour ride south of Glacier Park. I make my pilgrimage usually in August when the wildflowers are prime, but some years also in early summer, just after Going-to-the-Sun Road has been plowed open over the top of Logan Pass and melting snows cascade down every greening mountainside.

This year Bill and I entered at West Glacier and ascended through cedar forests and piney ridges to the Continental Divide at Logan Pass. The western heights were effulgent—a waterfall fantasia. But bear grass and wildflowers on the Highline Trails of the Garden Wall were just emerging from the sun-warmed earth and wouldn't reach full glory for a few weeks.

Having crested the pass, we descended toward sunrise country, winding past Saint Mary Lake and slowing at Two Dog Flats, where aspen groves glinted in the afternoon wind. We left the park at its eastern entrance at Saint Mary and continued north onto the windswept plains of the Blackfeet Reservation on our route to Many Glaciers Hotel.

Although I often go to Glacier in summer, autumn may be the most ecstatic season, with yellows of cottonwoods aglitter in the blue air. The brush is red leaved in September, edged with frost most mornings, and huckleberries are ripe in the grizzly bear haunts. For solitude, I go in winter. You can ski cross-country in a black-and-white landscape, the silence broken only by crows. Roads are blocked with snow. Tourists are gone, hotels closed, bears denned, and wolves on the prowl. But no matter what season, I love to walk Glacier's high-country trails, for at tree line the world explodes in a multitude of natural forms. It is a center of energy, the meeting place of two ecosystems—moist western forests joining and mingling with the more arid, rain-shadowed redrock country that runs along the east side of the divide.

On my first backcountry trip I went in mid-August with Bill and Alex and Andrew, then twelve. We set out from McDonald Lake up the six-mile trail to Sperry Chalet. It was hot as we hiked through the fern-floored forest of western red cedar and hemlock, then up three thousand feet into a region marked by subalpine fir and Engelmann spruce—the dwarfed, wind-bent, frost-stunted forms called krummholz. *Krummholz* is a word I love. Say it out loud. When you reach krummholz, you've climbed above six thousand feet.

Sperry Chalet sits like a mirage on a cleft of rock. For a few years it was closed for restoration and an environmental upgrade of sewage facilities, but it has reopened (thank the Lord and the park service) and is once again offering bunk-beds and home-cooked meals. On that long-ago evening in August we were thankful for shelter. The weather turned, as it will in these mountains, all of a sudden, and bad. We went to sleep on a summer breeze and awoke to six inches of snow, so much snow and fog that the four-mile trail to Sperry Glacier was closed. We put on all the clothes we'd brought, had a snowball fight, then began our trek down. By the time we reached Lake McDonald Lodge, it was sunny again. The boys went swimming while Bill and I sat on the deck in rocking chairs, sipping our gin and tonics.

Another summer, on a blue-sky morning, our same little family group climbed through ground-hugging alpine flowers, across terraced pools and hanging gardens, over a rockslide, and onto the ice of Sperry Glacier. In the bright air, surrounded by fields of crystallized snow, we felt on top of the world. Pulling ham sandwiches and apples from our packs, we sipped at bottled water. Bill reached for a pale object at his feet. He held up a used condom. Paradise was lost, at least for a moment (for some couple, we hoped, it had been found). We laughed. Reality. The glaciers of our longing are dedicated to wildness—the park, however, was meant to be (and is) a pleasuring ground.

The Kutenai, Flathead, and Kalispel tribes traversed Glacier Park while hunting, or for ceremonial reasons, but mostly as a path from the western forests to buffalo-hunting grounds on the plains. Blackfeet clans fiercely guarded the front, so passage was often a bloody affair. Only one small band of Assiniboine made their home in the mountains. Called Stoney because they used hot stones for cooking, these people barely survived. Stoney Indian Pass and Stoney Indian Lake in the park's northeast corner are named for them.

Then, from 1700 to the mid-1800s, only a handful of French, English, and Yankee trappers, fur traders, and explorers entered the region. A natural highway across Marias Pass at the park's southern border (at 5,216 feet, one of the lowest passes across the Rockies) was the Indians' secret. But as the United States expanded westward during and after the Civil War, the Blackfeet were decimated by smallpox, massacred by the U.S. Army, and—

after the last bison were slaughtered in 1882—starved into submission on a reservation. The gates opened.

Adventurers trooped into the mountains seeking precious metals, timber, and oil. Then came settlers wanting land, and sportsmen to hunt the game. It was exploitation, no romance about it, but the land was too harsh and remote. James J. Hill provided access with his Northern Pacific Railroad, which crossed Marias Pass in 1891. Hill, the "Empire Builder," realized tourism would be the only way to profit from such a place. After Glacier became a national park in 1910, he promoted the fabulous scenery and built great hotels to house visitors.

Travel in those early years was a more daunting enterprise than getting into your motor home or flying from New York to Kalispell, then renting an air-conditioned van. Voyagers stepped out of their sleeping cars near the park's southeast entrance at East Glacier. They were greeted by Blackfeet Indians who had been hired to entertain with drumming and dances. They spent their first night across from the station at Glacier Park Lodge, opened in 1914. Its lobby was (and still is) lined with gigantic pillars of Douglas fir.

From East Glacier tourists set off on horseback journeys across the web of trails lacing the park. Led by guides, these visits could last a month. How I wish I could experience such a trip. I see myself garbed in jodhpurs and fine leather boots like Amelia Earhart. I'd camp in tented enclaves or backcountry chalets. My meals of fresh game, exotically prepared by Chinese cooks, would be served on white linen with silver tableware.

* * *

Tourists can still take horseback day trips from outfitters at Apgar, Lake McDonald Lodge, and Many Glaier Hotel. We can ride narrow, winding Going-to-the-Sun Road (completed in 1932 by the Civilian Conservation Corps) in vintage open-roofed red Mercedes motor coaches, or in an air-conditioned bus piloted by a Blackfeet guide. There are sixteen roadside campgrounds in Glacier and Waterton where you can rub shoulders with fellow campers. But to touch the wild, you must leave the roads, put on your boots, and walk.

My favorite short hike is the boardwalk trail, a mile and a half round trip from Logan Pass to Hidden Lake. It's crowded but gorgeous, crossing tundra thick with alpine flowers and ending with an overview of Hidden Lake backed by Bearhat Mountain. There are marmots and ptarmigan along the trail, and always mountain goats on cliffs above the lake. Shaggy white and bearded, with black horns and dainty hooves, they seem to have just emerged from the painted mists of an ancient Chinese scroll.

The best place to see mountain goats is Gunsight Pass, 6,946 feet in elevation, about midway on the twenty-mile trail from Jackson Glacier Overlook to Lake McDonald. When we still did such things, Bill and I went backpacking there with friends from Seattle. At Lake Ellen Wilson, hot and footsore from the steep descent, Pat and I splashed in the icy waters. Suddenly we heard a shriek. Our friend Dave Nash had been snoozing on a rock. He felt something sandpaper rough cross his forehead and, opening his eyes, looked up into the beard of a billy goat. Goats crave salts. They follow humans, hoping for a lick of the minerals in urine. Or a sneaky kiss on a sweaty brow.

Animal stories are the best souvenirs you can bring back from Glacier. This is bear country, and both black bears and the hump-shouldered, dish-faced, dangerous, and endangered grizzlies may be spotted browsing along roads or in meadows like those below Granite Park. No one knows the exact population of grizzlies in the park; estimates range from eighty to two hundred or more. People love bear-attack tales, but the statistics are not so scary. From 1913 to 1998, there were 212 fatalities in Glacier Park: 48 drownings; 30 heart attacks; 25 vehicle accidents; 45 climbing or hiking fatalities; 4 horse-related mishaps; and only 10 deaths caused by bears.

Once I bushwhacked into the Apgars in fall, making a film about Doug Peacock, the legendary bear man. It was huckleberry season, and the hills were alive with grizzlies. We stood on a ridge observing a tarn where a Siamese-marked sow played splash-and-swim with her two cubs. Peacock knows it's best to observe griz through binoculars from a distance. But if you happen to encounter a grizzly close up, don't run, he says, for that will trigger its predator response. Avert your eyes, signaling submission. Stand tall, arms akimbo, speaking or singing in a calm voice. Back away slowly. If the bear charges, drop to ground and assume the fetal position, arms covering your head. Advice more easily given than followed.

A few summers ago I went to Waterton with my son Steve. We took the hour-long boat trip—a stunning ride along Waterton Lake, the blue Canadian Rockies rising on all sides. From Goat Haunt Landing we hiked south along Kootenai Creek about three miles to a deserted camp at Kootenai Lakes. Marshy, reed fringed, surrounded by sheer walls, this was ideal moose habitat. And sure enough, there stood great-racked Bull Moose, and Mom, and Baby—all knee deep in green water a few yards from our tent. We were awakened many times that night by the slurping, sighing, and groaning sounds that attended the moose family's feasting.

Next afternoon we were back in semicivilization, indulging in high tea at the Prince of Wales Hotel. This is another of James J. Hill's railroad fantasy tourist dreams come to full realization—a vision of Disneylands to come, but infinitely better because it is not virtual. Built in 1926, with steep green-gabled roofs and a tower, the hotel sits alone and magnificent on a bluff overlooking the lake. Our table faced floor-to-ceiling windows in the atrium lobby. The college girls who served us were dressed in red-plaid kilts. I wished for a bagpiper playing out on the deck at dusk, but what we got was a serious young woman plucking an Irish harp.

Bill and I ended our Glacier visit this year with a ritual return to Waterton. July 1 is Canada Day, and at the park's entrance we were let in free and given a small maple-leaf flag to wave. We'd planned to enjoy a very British eighteen holes at Waterton's *terrain de golf*, but I'd pulled a hamstring playing softball at my old co-rec team's twentieth reunion in Missoula and was too stove-up to do anything but hobble around or sit in a car. So we decided to see some new country and headed up the road to Redrock Canyon.

June had been a rainy month, and the lush meadows along Crendall Creek were emerald green, ignited here and there in colors red, white, and blue with lacy wild onion, scarlet paintbrush, penstemon, bluebells, and lupine. There were clumps of sticky wild geranium, Jacob's ladder, silky oxytrope, and white-plumed bear grass on the snow-streaked mountainsides that rose on all sides.

Heaven, I sang, head out the window, hair blowing in the fragrant breeze. *I'm in heaven* ...

PALO DURO CANYON

THE STAKED PLAINS OF TEXAS

Say you're driving the freeway south from Amarillo on the high, dusty plains of the Texas panhandle. The monochrome sky hangs darkly over the monochrome earth, and a storm-bearing norther stirs dust devils across the prairie. If you are me, you sigh. This is not the panhandle of your imagination.

I expected great herds of cattle. Horseback cowboys. Grass. A huge expanse of ocher-tipped swells bending to wind, like the plains Coronado disappeared into with his conquistadores in 1541. "I did not find their end anywhere I went," he reported, "plains with no more landmarks than as if we had been swallowed up in the sea. . . ."

Coronado's men marked their passage with stakes so they could find their way back, which is why these caprock plains are called the Llano Estacado. My mind is full of such lore as I drive across the rimrock plateau, and I don't appreciate the plowed and irrigated landscape of cotton fields, straight-line roads, and farmhouses bleached as colorless as the drifting skies. A helluva place to live, I think. But then the dun-colored surface cracks open and a great declivity cleaves the earth. It is a chasm brilliant with the colors of fire: Palo Duro Canyon.

At its widest this famous fissure spreads eight miles across; then it turns northwest, cut by Palo Duro Creek into a gorge only two miles wide from wall to wall. The erosion-resistant limestone caprock is sculpted into gigantic palisades mottled in eggshell whites. Softer layers of brick-red shales slope below the rim and are dotted silvery green with brush. Peering down from a viewpoint into the canyon's depths, I see badlands striped in pumpkin, orange and peach, darker siennas, and the lemony grit of sandstone.

Palo Duro Canyon State Park marks the entrance to the gorge. From here, a two-lane road runs down a steep grade past the Goodnight Riding Stables, Goodnight Trading Post, and Goodnight Dugout—all named in honor of Charles Goodnight, who pioneered the Goodnight-Loving cattle drover's trail to New Mexico in the 1870s and, with his partner John Adair, owned Palo Duro Canyon and ran a hundred thousand cattle over one million acres. *Goodnight* is a defining name in these parts, but if you ask me, the cowman's greatest deed was saving a herd of bison. The small wild herd that he corralled was said to be the last on the plains. Its members would become breeding stock that helped preserve the species in the United States. The iconic beasts we marvel at now all over the West are descended in part from Goodnight's chosen few.

No bison or cattle roam the park these days. It's a place to please people. You can bed down in campgrounds named for common flora in the region: Sagebrush, Hackberry, Sunflower, and Cactus. You can ride horses and mountain bikes or try the rugged nine-mile running trail along a high redrock ridge. And of course you can hike—in my opinion, the only way to see country. Midway down the gorge a maze of easy, sandy trails leads walkers up side canyons to Capitol Peak, Castle Peak, and the great eroded stack called Lighthouse, which is the park's landmark formation.

It is winter when I pass through Palo Duro on my way to Austin. Evening is coming fast and I cannot stay the night, much as I'd like to. But I can stop long enough to take a sunset walk. I can always stop to explore a luscious watering hole.

Water-carved walls enclose the box canyon. The air is soft and still as I amble past peach-colored cones and striped humps of gypsum. Junipers are heavy with berries, and I inhale their perfume. Then I notice heaps of fresh, berry-rich scat piled among the junipers and willows along the creek's mucky edges. Only a bear could defecate droppings of such amplitude. Suddenly, I'm alert. I did not expect a bear in Texas.

The wind kicks up. A cold, storm-driven gust nearly takes off my cap, makes me zip up my parka. The wind soughs like the unseen bear, reminding me of the ghosts who inhabit this canyon. The ghosts are Comanche—women, children, and old men trapped here on a September dawn in 1874—an unforgettable red dawn when U.S. Cavalry troops streamed down from the caprock, burned their villages, and slaughtered 1,048 Comanche horses.

Imagine the screams. The whites of turned-back eyes. The bloody, broken, writhing piles—horse on horse on horse. Soldiers ordered to shoot the animals are said to have turned away, puking in disgust. There was enough horror that day, enough slaughter of innocents, to break the spirits of hardcore Comanche warriors. Enough to end the Red River War.

I leave Palo Duro Canyon reaffirmed in my belief that there is no such thing as scenery. The beauty I encountered in that chasm is a mix of rosy clay and red blood, of white palisades and memory's sands. Palo Duro for me is conquistadores and Comanche and screaming horses and Charlie Goodnight's bison resurrected. It is a black bear who stands six feet tall on his hind legs, his breath rank and fur ruffled in a wind from the north. He is hidden in junipers along the trail by the creek, a mucky trail the colors of fire, which also holds my footprints.

FALLING INTO THE CANYON

It is the end of September, and for the second time in my sixty-some years, I am falling into the Grand Canyon. The first voyage was in May of '94—friends and lovers platooned on a large black motorized raft. The Colorado carried us down through amber, sienna, rose-colored layers of rock—a descent as vivid and unreal as dream. We fell into time and found bedrock, the schist called Vishnu glowing silver black, sculpted by water, older than life, two billion years old. I remember the trill of canyon wrens. A distant roar of rapids.

This morning, as the Canyon Explorations bus winds into the earth crack called Marble Canyon, I feel like crying out, "Hey, dream, here I come again." Alone. No love for anchor. Flame-striped mesas tower over me, and at my feet is the chasm. We are dropping through earth's crust toward a path that will be water.

To my delight, I discover that three of our guides are women: Nicole, a green-eyed Italian with a mermaid tattoo; Laura, who reads Mary Oliver's poems to us and looks like a model; and blond Meagan, the daredevil who runs up mountains. Larry, a veteran of the Colorado, will be our leader, and a wry artist named Sam will be the oarsman riding sweep.

Pulled up on the landing at Lees Ferry are four yellow oar-powered rafts loaded with gear and an empty six-foot paddle raft. This is the last day this season that motorized boats can put in, so there will be no gas fumes, no stutter of engines to break our illusion of wildness. We smear sunblock on our noses, adjust our not-quite-broken-in river sandals. It's a motley group. Not the well-heeled rich but working couples, brothers, families who saved sweaty cash to be here. Oddly, women are the driving force.

"It's our twentieth anniversary," says the Cape Cod lineman. "Linda said she was going with or without me."

"I'm just along for the ride," says a portly roadhouse owner from the Northwest. "She's the boss." His wife, a slim woman with big hair, smiles. What is it we women desire?

I, for one, desire always the physical. Stone, leaf, rain, blood. I want to see clearly the red-spotted toad, the blue beetle, the purple petals of a four-o'clock opening to sun. I want to scale cliffs, touch ancestral Puebloan shards. I want to wallow in the desert's heart. It pleases me that no one, man or woman, mentions the adrenaline rush of running rapids, the need to prove oneself in contest with the river.

Our first day in Marble Canyon we drop swiftly into rivertime. Hours are marked by the angle of sun, shadows on riffles, the growl of a stomach. The river cuts through strata of limestone, shale, and butter-gold sandstone, and we are as insignificant as the common white butterfly—a flutter of wings against ancient faces.

The Colorado, blue green at Lees Ferry, takes on the tone and texture of chocolate milk as silt-laden tributaries dump their loads into it. Even the turquoise waters of the Little Colorado, which emerge from sacred Hopi birth-springs, are running brown. It has been raining monsoons for weeks, and the canyon's slopes, ridges, gullies, and beaches are splashed with an improbable electric green.

The river bends, and I spot a blue heron etched on a cliff like an Egyptian hieroglyph. Another day snowy white egrets take wing over redrock. On warm stones a collared lizard with yellow feet pumps up on his hind legs like a mechanical toy. So much amazing life. In camp the night-blooming sacred datura unfurls its fragrant, hallucinogenic, and toxic white petals, beckoning the sphinx moth to feed and humans to dare the most dangerous sexual dreams.

At Pipe Creek I join the paddle-boat crew. It is day seven and we have put in just past Phantom Ranch, where our first week's passengers departed and a new group, footsore and sweaty from hiking down from the rim, makes ready to float the deepest canyons, the harshest rapids the river has to offer.

I feel like a veteran, having paddled two days despite a bum shoulder, and I'm showing off a bit for these dudes, but the main reason I choose to paddle is that I love the team effort, the action and focus, the breathless rush of riding square in the maw of the river. Our raft is strong, light, unencumbered by gear. Because it's more maneuverable, we can venture into rapids head-on, take routes the oar boats would avoid. Three rapids rated eight to ten on a scale of ten await us.

Meagan is at the helm. Paddling on either side are the new riders, all southerners, Della and her strong grown sons, a nurse and optometrist who have floated the river before, and me. We scream through Hance and Granite Rapids, stroking down glassy V-shaped tongues, skirting the holes, pulling through waves that rise above our heads. Meagan maneuvers us into the calm at the tail of each rapid and, released from danger, we let out a wolf howl of victory. She stands in the stern, eyes wild blue, hips swinging in her paddle dance. Living the moment like children, we raise our paddles above our heads, slam them on the water.

Meagan, at twenty-four, has run the river seventy times and never capsized. Now clouds darken the sky and the day turns cool. The rangers at Phantom Ranch had warned Larry a hurricane was approaching. Several people have been swept from side canyons by flash floods to drown in the river. A headwind blows in from the west. Only one rapid to go. Hermit, the roller-coaster ride Sam told me is the most fun of all.

First, as always, is the roar. "Forward," shouts Meagan. We drop from calm into havoc, attacking Hermit's wave train head-on. One, two, three, four giant waves lift us. All I can see is the high brown foam, the dense fury into which I must place my paddle as we rise and fall. Breakers slam into us, slap our faces, drench us in icy blasts, but we keep stroking to a beat set by the lead paddler—stroke, stroke, stroke—for a paddle wedged in water will keep you saddled on the raft's slick yellow sides, and momentum carries us through even this fierce water.

Then comes the fifth wave. We climb up and up, stroking hard, but we do not cut through. The wave grows. It's a demon curling above us. Its foaming dragon breath is as distinct as a Japanese painting, alive. Now I am stroking air. The raft is hanging vertical in air. Then the crest breaks. We are thrown back. The boat flips. We are flying.

I go down a long way. Buried in the brown deep, I'm a pebble in a maelstrom. This is power pure, power strong enough to drive turbines. I am holding my paddle. I let it go. I push up through the icy turbulence until breath leaves me and I am choking, swallowing water. A gulp of air, blessed air, then I am driven down in a whirlpool. Up and down and up again. The river roars past rock walls and I'm still helpless.

Inert as driftwood, I bob toward an eddy, held afloat by my life vest, dazed, numb, gulping air. And I'm drifting away from the rescuing oar boats. I swim toward Nicole's raft, kicking despite exhaustion. A man's hefty arm grabs mine. "Annick," says Nicole, "you're bleeding." She leans close, her long brown hair in my face. "Who am I? Do you know what's happening?"

"I'm okay."

She grabs my vest and yanks. It isn't tight enough. It slips above my eyes. "Can you give me some help?" she asks me.

I want to help, but I'm a stone. When they pull me in, I lie on the boat's self-bailing bottom, gasping like a landed fish.

The night is windy and cool, with no stars. My head throbs from a welt across my forehead, a cut below my right eyebrow, a bruise across the bridge of my nose. All the rest who had "gone swimming" are equally shaken, but not injured. It seems a paddle hit me as I went overboard. Maybe my own. Probably the handle of the paddle in front of me. Forsaking a mirror, I smear my wounds with oil of arnica, take two aspirin, drink a gin and tonic with the crew. I down a beer at dinner, swig brandy from my stash, pull on long underwear, and still I'm shivery. I put up my tent for the first time, glad for cover.

"Tonight," says Larry, "you may wake up crying. It's a thing that happens. Come over and I'll comfort you."

"I don't think so," I say. If I need comfort, I'll go to the river women—Nicole, Laura, or Meagan, who is also wounded, nursing her pride.

Deep in the night I have a nightmare vision. An amorphous brown glob like a giant amoeba advances toward my tent. It flows closer, relentless, unstoppable. I know it is coming. I wait. Its weight creeps over my head, my chest, crushing breath. I lie passive as the brown glob engulfs my legs, my feet. The terror seems endless, and then it's gone. I watch the dark thing creep away. I am awake, tingling with fear. I know it won't come back. I'm safe.

A few days later I ride fearsome Lava Falls in Nicole's boat, face forward into the waves to keep the prow down. She skirts the ledge, glides skillfully through the thirty-foot drop. Every craft passes the test and we let out a great cheer. On the thirteenth day I am confident enough, despite a creepy hangover of horror, to mount the steed once more, paddling front right, setting the pace through the final seven-plus rapid. It has taken four days to gather courage, but the ride is perfect and I am redeemed if not reborn.

Our last camp is on a hot beach at mile 220. We use the paddle raft as a slide, leaping and splashing into cold waters. I smile, thinking we look and act like a bunch of otters, sun browned and salty, joking and playing, the best of friends. The river has allowed us to pass in ecstasy and fear—men and women together as a team—and now we can laugh. At dawn we will float the final six miles in silence.

After dinner, in the warm dark, people talk about what this journey has meant. "It's made me stronger and more humble," I say. "More surefooted, but less sure in my head. I've learned a few things about mortality."

I am referring to Hermit Rapid, but also to a sunstruck hike up eighteen hundred feet to Thunder River. Falls thundered from limestone caverns down a rock face to a leafy oasis. There were hanging gardens of maidenhair fern and cardinal monkey flower, aquamarine pools. This, I felt, was a sacred source, a place I could return to again and again in memory.

And I am also referring to the bighorn ram who stood looking out at the canyon from the top of a high lava outcrop called the Gorilla's Head. Dusk was falling. It was our last evening on the river. Darkness washed into the U-shaped valley where we were camped, but up on the rims, sun gilded the ram's horns, which swept forward in heavy curls, like waves.

The ram stands at the apex of his world. He is my living shrine, transient and magnificent as the canyon wren, the Bright Angel shale, the fire ants, and us. He tells me this path down the river is not a fall into dream but into the great reality—the only one and only.

PART 4

BIG BLUESTEM

I. Begin with Grass

I believe a leaf of grass is no less than the journey-work of the stars.

Walt Whitman, from *Leaves of Grass*

My first sight of the Osage grasslands is in late August, when seed heads droop heavy and the lush green of summer has begun its fade to colors more earthen. The hot, humid air reminds me of Chicago—the "prairie city" with no prairie in sight—where "fog creeps in on little cat's feet." Chicago, my childhood home, was celebrated by the poet Carl Sandburg, who also sang the older songs of grass. "The prairie sings to me in the forenoon," wrote Sandburg, "and I know in the night I rest easy in the prairie arms, on the prairie heart."

Sandburg's Illinois, and mine, was tallgrass once, a broad stretch of big bluestem like the Oklahoma prairies of Osage County. But the Osage does not have the rich, deep, tillable soil of Illinois. It grows cattle, not corn, and beneath its skin of limestone and sandstone lie pools of oil and pockets of gas. Driving the sixteen miles of gravel road from Pawhuska to The Nature Conservancy's Tallgrass Prairie Preserve, I feel spooked. I am a stranger in a land I have never seen, know nothing about, and am as bewitched and bewildered as the pioneer women who arrived on the Kansas and Oklahoma frontiers to claim Indian lands more than a century before.

"Lost! Out here on this lonely prairie," wailed Martha Lick Wooden on a June night in 1878. Martha's husband had missed the wagon trail to their Kansas homestead, and the howls of coyotes seemed to her like hyenas—wild dogs laughing at the imbeciles who could not find their way in a wilderness of grass.

"It was such a new world, reaching to the far horizon without break of tree or chimney stack; just sky and grass and grass and sky. . . ." Lydia Murphy Toothaker said, recalling her first night on the prairie in 1859. "The hush was so loud."

Lillie Marcks was seven when she rode into the grasslands. "As we drove on the prairie, Mother and I could hardly stay in the wagon. The wild flowers covered the prairies in a riot of colours like a beautiful rug. How we longed to gather some."

The Oklahoma prairies that run on all sides of my rented Mazda partake of the West. Broad horizons where grass meets sky are western. I feel at home in windy blue spaces, but this is not the dry, light, high plains I love, or the mountain meadow of my homestead ranch, swept by breezes even in the dog days of August. This is borderland, a place where West begins.

Always the hunting ground of nomads—a grazer's paradise—the Osage prairies once thronged with bison and elk, antelope and prairie chickens. Now many native animals have been hunted out, replaced by cattle and horses. I know there are trails in the deep grass made by Indians, explorers, ranchers, and oilmen. And if I looked closely, I would notice that the land is littered with decaying oil pumps, house foundations, rutted tracks. But I am inside the big picture today, overwhelmed with undulating prairies that sweep to every horizon. To my newcomer's eye, the immense landscape shows few signs of human habitation or industry. I see only grass, a few lone post oaks, blackjack thickets on the perimeter, and sky, sky, sky.

Urban civilizations, from their origins along the Euphrates, have taken hold around places where agriculture and irrigation offer stable sources of food, or at the crossing of major trade routes. But where the land is so intractable that people can only travel through it, dig holes into it, or trail their herds over it, the landscape remains more or less intact. This is the prairie where soil is so thin and rocky it could not be plowed and farmed, so this is the prairie that survives.

When I stop my car along the gravel road, I catch scents of musk and dust and rotting stems, an electric charge of ozone. The Oklahoma air is warm, moist, and oppressive. Southern. All I see is grass and more grass, nothing but grass that, when I step into it, envelops me. Claustrophobia panics me, and I push the coarse, enclosing leaves away from my face. I cannot anchor myself to the soil at my feet or the wind at my back through connections of memory or experience. I feel the sky descending.

In the fall of 1832, just back from a long sojourn in Europe, Washington Irving set out on an expedition to see the wild prairies of Indian Territory. He was anxious to reroot himself in the American experience, and what could be more American than the western frontier? In *A Tour on the Prairies* Irving described what he saw. He was standing on land not far from where I stand, perhaps, like me, watching the grasses shake their heads in the rustle before a storm.

> *To one unaccustomed to it, there is something inexpressibly lonely in the solitude of a prairie. The loneliness of a forest seems nothing to it. There the view is shut in by trees, and the imagination is left free to picture some livelier scene beyond. But here we have an immense extent of landscape without a sign of human existence. We have the consciousness of being far, far beyond the bounds of human habitation; we feel as if moving in the midst of a desert world.*

The green prairies all around me speak of a fecundity no desert ever has, yet in an emotional sense Irving was right. A land seems desolate if you know nothing about its geology, botany, and biology, know no histories or stories. Even in my ignorance I know that this place, like every place, holds a history of habitation by birds, insects, small and large mammals (including humans), and that this prairie's story is underpinned by tall grasses and flowering plants—the roots beneath my feet.

If I could become familiar with the meadows and gullies, the peninsulas of post oaks and blackjacks, the trickle of streams over outcropped limestone, I would not be lonely. The red-tailed hawk riding air currents is not lonely, nor the spider spinning its gossamer. To know the stories of this place, I must begin with grass. To understand grass, I will have to descend to its roots, and beneath roots to soil and bedrock. But first, as always, is the sky.

"I believe a leaf of grass is no less than the journey-work of the stars," wrote Walt Whitman, America's great poet of grass. The evolution of the tallgrass prairie may be thought of as beginning in stars, in the forces that created the landforms of our continent and will continue their work long after our species has become embedded in rock like dinosaurs and trilobites.

Once upon a time, say sixty-five million years ago during the Cretaceous period, there were no Rocky Mountains and no great American grasslands either. North America was a level, forested country from sea to shining sea. Its climate was warmer and wetter than today's, with no deep-freeze winters or steaming summers. Then the bedrock began to heave and lift under pressures of clashing tectonic plates. During the Laramide Revolution the continent's western edges rose up in great bulging faults and folds, which eventually became the Rocky Mountains, the Sierra Nevada, and the Sierra Madre of Mexico.

Prevailing winds, which have always come from the west bearing Pacific moisture, ran up against the new mountains. Instead of dropping their rain burdens evenly across an even landscape, the air masses were forced upward over peaks and ridges. Clouds cooled as they rose, condensed, then let go of their moisture in great falls of snow and rain on the westerly slopes of each range. By the time the Pacific westerlies reached the Continental Divide, they were exhausted of vapors.

The result was what we call a rain shadow. In the dry shadow on the eastern side of the Rockies, about twenty-five million years ago, the great North American grasslands began. For eons they seeded their slow way eastward, aridity and wind helping grass devour forests in a continual movement across the continent.

In addition to the drying effect of the rain shadow, a climate shift to warmer, drier air opened holes in the overstory of weakened forests. The sun shone where it had never shone before. Pines gave way to more drought-tolerant hardwoods, which in turn gave way to grass. Trees could not survive as well as the deep-rooted grasses in increasingly extreme temperatures and under the pressure of high winds. Winds blew seeds and spread lightning-caused fires. Fires opened new areas.

The original North American mastodons, mammoths, and ground sloths evolved into gigantic creatures in order to browse brush and lower branches of trees—keeping the forests open. Grass invaded the open regions. Smaller mammals, including horses, camels, bison, and swift antelope, evolved hooved feet, hypsodont teeth, and digestive systems designed to take advantage of the grassland environment. Carnivores preyed upon the grass eaters in a cyclic balance of eat-and-be-eaten, but there were enough grazing animals to keep fuel sources scarce, diminishing the destructive force of fires and increasing diversity.

The grass kept growing eastward beyond the edges of the rain shadow to a region where westerlies collided with the warm, wet currents of air blown inland from the Gulf of Mexico and southern seas. This created the turbulent climate where tornadoes flourish and the tallgrass thrives near borders of forestlands.

Climate remains the crucial force in creating prairies, but nothing stays static on this earth for long, especially climate. Long dry millennia alternated with ice ages. Glaciers slipped down from the north in powerful rivers of ice. They scooped and leveled the northern plains, creating glacial soils that later, blown into drifts by the constant wind, became the rich deep midwestern soil called loess. Although dry spells stopped the glaciers before they reached southern Kansas and Oklahoma, the influence of ice traveled south with seeds carried by wind and birds, and on the hides of migrating animals.

The Osage tallgrass prairie is only about eight thousand years old. It lies in the southern portion of what was, until only a couple of hundred years ago, a huge contiguous grasslands ecosystem of about 142 million acres. Think of a picture of North America taken from outer space in 1803, when Jefferson made his Louisiana Purchase. Focusing in on the continent's center, you would see the Osage as the index finger of an immense deep green hand, a stripe of tallgrass that began in Manitoba and ran south more than one thousand miles to the Gulf of Mexico. At their widest, the tallgrass prairies extended from western Indiana into Nebraska.

Continuing to trace a west-moving line across the continent, the grasslands become shorter—melding in my imaginary map into a pea-green second stripe called the midgrass plains, which hooked south from Saskatchewan through the Dakotas to Texas. At the hundredth meridian, where the high, dry climate can't support grasses dependent on abundant moisture, the color pales to the gray-green of bunchgrass, which is dominant on the shortgrass plains running south from Alberta to New Mexico, then west to the Rockies.

When I flew from Missoula to Tulsa, I looked down on a land where the grassy once-great plains had been replaced with checkerboards and circles of corn and wheat. Hundreds of towns lit the night, their yellow lights like stars roosting, and I rested my head on the window, shocked with the recognition that this huge transformed landscape was not much older than I am.

In contrast, when I finally reached the Osage prairies, the grasslands appeared primeval. Impenetrable. I saw myself standing apart from the landscape. Dark stands of post oak and blackjack seemed to be pushing their way into the prairie. The muted land was feather soft and cushioned. A wind at my tail stirred plumes of big bluestem and switchgrass in the dipping and rising folds of great meadows. Clouds flew over the wind-tossed grass, casting deep purple shadows on silvery greens and grays. Below the roof of the advancing storm, turkey vultures circled with black, outspread wings.

The main building of the Tallgrass Prairie Preserve is a Texas-style one-story redbrick, U-shaped structure built in 1920 and called the Bunkhouse or Headquarters. Before The Nature Conservancy bought the Chapman-Barnard Ranch from the Barnard trusts and refitted the old ranchhouse for organizational uses, it was an actual bunkhouse as well as the ranch's headquarters.

In its prime the 125,000-acre ranch was one of the great cattle empires of the region. When Chapman died, the ranch was divided. The Conservancy's preserve includes nearly thirty thousand acres from the Barnard portion, as well as several thousand acres of additional property. I stand on the original, arcaded front porch of the bunkhouse with Mary Barnard Lawrence as dusk turns oaks and prairies into a quilt of muted greens and browns. Mary is in her seventies, a tall, elegant ranchwoman. She is the Barnard heir whose desire to turn the old ranch into a symbol of the living prairie made the conservation effort possible. Mary, in her long, divided skirt and crisp white blouse, is, it seems to me, the generous heart behind the preserve. She leads me to a grassy mound where we can watch the progress of a storm sweeping from the west.

A clap of thunder rolls toward us like timpani, and lightning streaks the sky iridescent blue. A great rain begins to fall. I feel like taking off my shoes and leaping into puddles like my sisters and I used to do during the warm, summer storms of a Midwest girlhood.

After the storm, in the misty light of a blue moon, cicadas begin a strident chorus in a hundred-year-old hackberry tree that dominates the courtyard outside the Bunkhouse porch. I am not used to the voce of this insect, its shrill buzz of life. I want to know the hackberry's secrets, to be the bug in its branches. Then, under the croak and buzz of cicada wings, I hear a muffled, thumping sound.

It's the ghost drums of Osage Indians, I think. The hairs on my arms rise. "We are here, where we belong: boom, boom," the drums say. "Who the hell are you?"

Next morning I go out into the fields with Bob Hamilton, the preserve's scientific director and ranch manager. Bob is a wildlife biologist, a Kansas prairie boy—literally boyish in his checked shirt, jeans, and brown tousled hair. He laughingly informs me that the rhythmic pulse that startled me last night is only an oil pump on the prairie, giving voice to an oil-producing presence that, like it or not, is still here.

We walk into the tallgrass under a sky of pure blue with powder-puff clouds skittering like sheep across the pastures of heaven. In the wake of the storm, the temperature dropped during the night, taking summer away with the heat. In a space of twelve hours the prairie has turned toward autumn. Amid the green meadows I notice splotches where bunches of big bluestem have darkened from the rain and the cold. By October, the whole landscape will turn russet and brown. In winter it will be drained of color, white with snow. Not until the young shoots spring up next April will green return.

Bob leads me into a stand of grasses that rise from ankle height to stems almost ten feet tall. As I pick a bouquet, he names the grasses.

"Big bluestem. That's the tallest. You can tell it by the three-part seed head. Looks something like a turkey's claw. That's why some people call it turkey-foot bluestem."

Big bluestem is the king of tallgrass—a native, warm-season perennial. It loves moisture and good soil and thrives in valley bottomlands. Under our feet the sod is a massed web of big bluestem roots, which can reach down almost a dozen feet.

Bob Hamilton speaks of grass with the pride and authority of a father praising his prodigal children. He tells me that the underground vitality of species such as big bluestem—their deep and pervasive root systems—enables them to survive drought, fire, subzero temperatures, even intensive grazing. Like other prairie grasses, big bluestem sows its seeds on the wind and also spreads by rhizomes. Its discarded, curled leaves provide a protective mulch that creates new topsoil as well as fuel for wildfires, which are crucial to prairie evolution.

Bob reaches down and picks a delicately fringed stalk from a knee-high bunch of grass. "This is little bluestem," he says. "It's one of our dominant warm-season grasses, along with switchgrass and Indian grass."

I learn that little bluestem was the most abundant grass of the American prairies, native to every state except California, Nevada, Oregon, and Washington.

There's a smatter of little bluestem on the borders of my Montana meadow, and in small patches throughout the Midwest. One of the most nutritious and abundant of all prairie grasses, little bluestem was the wild hay that pioneers cut and stacked as they settled the West.

High as the bison logo on my baseball cap, the prettiest tallgrass within reach of my gathering hands has a perfect plume of yellow seed heads at its top and foot-long leaves spreading at forty-five-degree angles from its stalk. I would love to pick a great armload of this grass and stand it in a vase beside my fireplace.

"That's Indian grass," says Bob. He points to the claw-like ligule that emerges from the stem wherever a long leaf grows. "It's one of the most nutritious grasses on the prairie."

Before I am finished, there will be a host of other grasses in my wild bouquet, with common names so descriptive you can imagine what they look like: *switchgrass, foxtail, blue grama, hairy grama, windmill grass, sand dropseed, purpletop.* The names are poetry. Knowing the names, I am no longer claustrophobic. I believe I could feel at home in the tallgrass, like bobwhite quail, or blue dragonflies that hover amid spiky leaves, or the common garter snake that wisps away from my boots, a creature almost indistinguishable from the earth.

There is more in the tallgrass than grass. There is a splendor of color. I point to the understory, where August's muddy greens are made bright with yellow petals and blue rosettes.

"Wildflowers!"

"Forbs," says Bob, using the more specific name for broad-leaved plants. "And that one you're picking, it's a legume."

Any gardener should know a pea when she sees one—the dark green opposing leaves like feathers on a stem. The cupped blooms that will fruit into pods. The legume that I add to my bouquet is called showy partridge pea, because of its hot yellow flowers.

Legumes are a natural fertilizer, returning nitrogen to soils, and many species, including clovers, are native to the prairies. Tick clover has hairy seedpods that cling to clothing or hair like a tick; wild alfalfa is a nutritious June-flowering plant with tiny purple blossoms; purple prairie clover rises with spring grass and is high in protein; roundhead lespedeza sports fine silvery hair and is topped by soft golden flower heads; and the accurately named wild-blue indigo is slightly toxic, but its sap turns purple when exposed to air.

Also yellow, but taller than the legumes, is goldenrod, which I thought made me sneeze. But I am wrong, for goldenrod is pollinated by insects, not by pollen-spreading winds that infect some of us with hay fever. A tea made of goldenrod warmed prairie explorers. Henry Ellsworth, U.S. Indian Commissioner for the West, served some to Washington Irving during their expedition to Oklahoma in 1832. "—it is *sudorific, gently stimulating* and an active diuretic—in large quantities it is laxative," wrote Ellsworth. "Mr. Irving is so much pleased with it, that he has ordered a quantity for New York."

Goldenrod has gleaned its bad reputation by consorting with the extremely sneezable ragweeds. Bob leads me to another place on the prairie, a spot of scorched earth dominated by forbs rather than grass. Western ragweed grows in rough, bushy clumps on all sides of me. I notice more goldenrod and prairie dropseed, and lots of tall, aptly named broomweed, whose open-fingered seed heads look like upended umbrellas.

The Conservancy burned this piece of ground during the summer to mimic the seasonality of historic prairie fires in the stretch of time before settlement, when herds of native bison would devour the young bluestem shoots that sprang up in fire's wake. They reintroduced not only seasonal fires but also the bison, starting with three hundred head and aiming for two thousand as the herd reproduces. Eventually, using these ancient means of prairie incitement, they hope to re-create a grasslands environment similar to what the first European explorers saw.

But for generations, summer burning has been an unheard-of practice for cattle ranchers, who wouldn't dream of destroying their high grass. After the burn, local ranch folks drove their pickups along the preserve's gravel road. On one side they saw thick grass; on the burned side, a weed patch. "Look at the weeds!" they exclaimed, confirmed in their suspicion that the preserve's managers are out of their minds.

The notion of "weeds" as demonic forces to be destroyed is a value judgment outside of nature's order. Agriculturists despise the rugged invading armies of "pioneer" plants such as ragweed that rise up where grasslands have been heavily disturbed by fire or grazing. Such forbs have a role in a healthy ecosystem but hold no nutrition for people or domesticated animals, and are therefore deemed useless. Ragweed spreads rapidly by means of its long underground stems. Cattle don't like it much, and if a dairy cow eats ragweed her milk will have a bitter taste.

Ordinarily we fight such weeds with herbicides. But on this preserve native plants are welcomed (foreign invaders with no natural enemies are another matter). Bob Hamilton is a patient manager. He knows that in time the native grasses will return with no need for human interference, fulfilling their old patterns of growth and diversity—a changing mosaic in which grass is only one element. A natural prairie, he explains, operates with different rules than a cultivated garden or monoculture. The cyclic movements of life forms throughout the ecosystem is necessary for its health and diversity. The bison that grazed this summer's burned patch have already moved on to better pastures, and next year the "Big Four" grasses—big bluestem, little bluestem, Indian grass, and switchgrass—will begin their comeback. By the second year, if there is no fire, this stand will hold more grass than weeds; and by year three, says Bob with a hopeful grin, "the late-successional grasses will triumph."

My attention is diverted by a scattering of delicate blue-flowered pitcher sage and a few red-purple bunched heads of tall, stalky ironweed. I scan the meadow for the rosy blooms of blazing star or button snakeroot, but find none.

"I thought there'd be compass plant," I say, looking in vain for the giant stalks with yellow sunflowery blooms.

A clan of Osage Indians who called themselves Walkers-in-the-Mist used seven-foot-tall compass plants to plot their routes across fog-bound prairies. Wagon-train scouts marked trails for their followers by tying flags to the flowers' stalks. I would love to see the foot-long oak-shaped leaves that give this characteristic tallgrass prairie forb its name—leaves whose long sides point north and south, and whose flat horizontal edges face the rising or setting sun.

Bob shakes his head a bit sadly, for compass plant is rarely seen on fenced pastures and meadows that, for generations, have been heavily grazed by cattle. Many typical wildflowers have grown scarce on the prairies, or have disappeared, because—unlike bison—cattle devour broad-leaved plants along with grass.

"About 75 to 80 percent of a cow's diet is grass, the rest forbs," Bob explains. The Conservancy hopes to bring a more natural distribution back to the grasslands by replacing cattle with bison. "At least 90 percent of a bison's feed is grass alone," Bob continues. "In our bison pastures we expect to see a resurgence of more broad-leaved plants.

"Agriculture," Bob continues, "has resulted in a disconnection of space and time. With our introduction of fire and bison, more compass plant has come back."

Some native plant species that were grazed heavily or killed with herbicides spread by previous ranchers may be scarce now, but ther are clusters of surviving forbs in remote byways where cattle have not grazed. Their seeds will return on prairie winds or be carried by birds. Other seeds are storing their energies under the sod. Given the right conditions, the wild garden is returning.

"When we made our plant count on the preserve," he continues,"we found 650 species.They were all here to begin with.We don't expect any new species to turn up. But the abundance and distribution will change.We are trying to get comfortable with change," says Bob. Not an easy trick.

I find myself daydreaming. I imagine a host of many-colored flowers that sleep in underground beds, nurturing seeds like bears nursing cubs in the dens of hibernation.

As we walk toward the Bunkhouse, Bob points out soils that underlie the landscape. To the west are limestone-based soils, the most productive for tallgrass. I see bluestem meadows edged by a fringe of brush and oaks. Looking east, I see a cross-timber forest of blackjack and post oaks that juts into the grasslands. Those tree-growing soils are sandstone based, says Bob, often exposed, thin, and rocky. Then we look down at some loamy creek bottoms, which, he says, is the only kind of land that can be farmed here—land where grasses grow ten feet tall.

"The Conservancy's policy," Bob says as we enter his book-lined office, "is to reintroduce randomness" in order to encourage biodiversity. It's a paradox, I tell him—trying to manage this natural prairie using nature's methods, which are by definition unmanageable. "We're only trying to restart the engine," he adds, "trying to mimic what we think is a functional landscape."

On the way to my room I rest in the shade of the old hackberry tree by the Bunkhouse porch.The grass here is clipped short, a green undifferentiated lawn—a place where you could picnic in comfort, play croquet, spread a blanket and read a book.This is what most folks think is a functional landscape.

My arms are still grasping the huge bouquet of grasses and flowers I have brought back from the prairie.What will I do with this sloppy bunch of stems and leaves already shedding seeds and petals? It is too fragile to pack in a bag, wrap in plastic, stow in the narrow compartment above an airline seat. I cannot hold on to this bunch of grass except in my mind and memory. And that is where the landscape I have just traversed becomes most functional for me. The new landscape in my mind has opened my imagination to a world of grass dizzy with motion, form and color. I am ecstatic, addicted, and I am not alone in this.

Rolla Clymer was a newspaperman in El Dorado, Kansas, who traveled through the Flint Hills for thirty-nine years on his way to and from Topeka. He always loved the prairies but did not write his regionally famous editorials about the land of tallgrass until he was an old man, retired and reflecting. At the age of eighty-five, voiceless from surgery, he continued to put pen to paper in rhapsodic paeans to his great obsession and joy. "Now, as midsummer nears," he wrote, while his own life waned:

> *The same sweet grasses bend their heads before the South Wind's vagrant whim; the same riot of wildflowers runs glitteringly through the ex-*

> *panse; the same anthems of birdsong rise from feathered throats; and the same smoky haze dances afar off where the sky bends downward to earth. Over all this comely region the same great solitude prevails—a silence that is its distinctive symbol—a reflective calm with healing in its wings.*

I promise myself to return to the prairie when the great wildflower show takes place with sulphur butterflies and monarch butterflies, hummingbirds, daisies, larkspurs, purple coneflowers, and fragrant wild roses. The new grass will be emerald, thick as the carpet of a Persian queen, and I will lie down in it. I recall two favorite lines from Andrew Marvell's poem "The Garden"—"ensnared with flowers. . . /. . .I fall on grass."

"I would be converted to a religion of grass," wrote Louise Erdrich in an essay about the prairies of her Ojibwa ancestors near Wahpeton, North Dakota. Now there's a sensible religion. We should be baptized in grass. Grass, our nest. Grass our spiritual home.

"All flesh is grass," says the prophet Isaiah in the Bible. My nostrils are filled with the sweet scent of grass. The tingle of grass brushes my hands. The itch of grass is in my eyes. I am tempted to sing "Yes" to the grass. All flesh is one.

II. Dancing Back the Buffalo

> *The whole world is coming,*
> *A nation is coming, a nation is coming,*
> *The Eagle has brought the message to the tribe.*

The father says so, the father says so.
Over the whole earth they are coming.
The buffalo are coming, the buffalo are coming....
Plains Ghost Dance song, 1890s

In the bison pens of the Ken-Ada Ranch just outside Bartlesville, Oklahoma, the ground has been eaten bare. Dust blows behind the wheels of the all-terrain vehicle that transports us toward the shaggy, rather pitiful beasts. We stop at a safe distance and proceed on foot. One bull raises his massive horned head, bends his forelegs under him, and rises from the dirt. He gives us a spiritless lookover and ambles away.

The bison cows seem in worse shape. Many are rib thin, their flanks marked with red, open wounds. Bulls nip cows in the rough game of mating, my guide Maryan Smith explains, and at least some of the sores have arrived via sexual encounters. Which is what we are looking for as we trail through the bison enclosure following a film crew from the *National Geographic.*

Maryan is the film's director. "It's supposed to be rutting season," she says, giving her long brown hair a shake and a toss. "We've been waiting and waiting all month . . ."

"Maybe they're camera-shy," I say. It occurs to me that the sexual energies of these animals must be low because they're depressed from being crowded together in these desolate pens. Then again, perhaps like us they prefer to do it in privacy.

Maryan wants to record the dramatic summer rutting show that bulls enact in the wild. The bulls, however, are biding their time and gathering their energies. "There is perhaps not an animal that roams in this or any other country," wrote the fur trader Alexander Ross, "more fierce and forbidding than a buffalo bull during the rutting season."

To win his desired cow, a bull has to fight off rivals, courting her for days, tending her. Challenged, he will raise his plumed wisp of a tail, which engorges for this purpose. His hair bristles, his eyes bulge, and he utters guttural roars. Bulls in rut face each other, paw the earth, dig their horns into sod and toss up chunks of grass.

If bluff doesn't work, a courting bull charges, head to head, trying to bring an opponent to his knees. Only the mat of hair and thickness of bone beneath his brow keeps his skull from cracking on impact. If one bull falls, the other may rip his flank with sharp-pointed horns. A prime bull, say eight years old, can do fatal damage to an old giant. It is better for the oldster to turn tail and run.

During rutting season a bull may lose two hundred pounds. His bellow becomes a whisper. He sulks, pants, drags his great blue tongue through the grass. Meanwhile, the cow remains calm, holds her weight, and when she is ready the mating takes only a few passionless seconds.

Every American knows what a bison looks like, but bison in the flesh are another matter. Imagine seeing a wild herd for the first time. The Spanish explorer Captain Vincente de Zaldivar described the legendary buffalo of the New World: "Its shape and form are so marvelous and laughable or frightful, that the more one sees it the more one desires to see it, and no one could be so melancholy that if he were to see it a hundred times a day he could keep from laughing heartily at it many times, or could fail to marvel at the sight of so ferocious an animal."

Bison are the only bovine species with long hair—a thick, curling, dark brown shoulder cloak. A mature bull stands more than six feet tall and can weigh nearly a ton. Cows are smaller. Both sexes have convex foreheads, shoulder humps, and heavy, curved horns, although the bull's horns are larger. The animals appear top-heavy and ungainly, but bison are equipped to survive in mountains as well as lowlands, in forests as well as prairies. They prefer grass yet will forage on brush; they can subsist for days without water while moving to wetter ground. Bison face into storms, their heavy coats enabling them to weather extreme cold. When summer comes, they shed fur in ragged molting patterns. They run faster than most horses, are agile enough to scale heights with mountain sheep, and can jump over obstacles up to six feet high—standard height for bison fences.

The original, extinct American species, *Bison antiquus,* ranged from Alaska to Nicaragua, and from eastern woodlands to Oregon and California. They evolved into the species we know as buffalo or *Bison bison,* although some taxonomists have replaced the genus name *Bison* with *Bos*—the same genus as cattle. There were two varieties in North America, the woods bison in northern Canada, which are larger and darker than the predominant plains bison, and have shorter beards adapted to brushy habitats.

Plains bison populations grew to astronomical numbers: Estimates run from thirty to sixty millon. Nomadic herds foraged tallgrass, midgrass, and shortgrass plains, but the largest groups gathered near the middle. There were many reasons for the population explosion, a prime one being longevity—bison in wild or semi-wild environments may live for fifteen to twenty years, with cows giving birth to one calf each spring.

Because of their bulk, strength, and fleetness, bison could defend themselves against all predators except armed humans or the occasional ravenous grizzly bear or mountain lion. Gray wolves culled the old, sick, and weak animals, keeping herds prime. And the symbiotic relationships of bison with grass and fire kept food sources plentiful, providing constant new range.

Before horses and rifles enabled Indians to be more heedless in hunting, and before their mass slaughter by white traders, settlers, and commercial riflemen, nature was the bison's worst enemy. Prairie fires could burn and blind the beasts, or diminish the sight of already weak eyes. Spring blizzards might freeze hundreds of calves, and tornadoes could toss a bison into oblivion, just as they toss trees and houses. During hard winters herds crossed ice-covered waters with no problem. But a bison cannot know from the banks that the ice ahead may be thin from spring thaw. Once leaders begin crossing a frozen river the herd will follow, even when ice gives way and pushes one after another under to drown. John McDonnell, a trader on the Qu'Appelle

River in Canada, watched 7,360 shaggy carcasses float past him one spring day before he tired of counting. Instinctual follow-the-leader behavior probably did more damage to bison than weather or fire.

Stampedes were also deadly. Any chance alarm could start the herd moving, and once in motion it gathered its own momentum. In 1541 Coronado's southwestern explorers reported seeing a stampede run into a ravine. Some animals fell, then more fell upon them until the gap was filled to the brim. And still the stampede roared forward, the main herd running over the bridge made by the trampled bodies beneath their hooves.

The three hundred bison I see gathered in the Ken-Ada Ranch enclosure are an anomaly. Cattle have dominated the few remaining natural prairies in Kansas, Oklahoma, and Texas since the 1870s, when the wild bison were hunted out, and the Ken-Ada Ranch is primarily a cattle operation with summer grazing leases for thousands of steers. But it also raises bison, and this bunch is headed for the Tallgrass Prairie Preserve, where its cows and bulls will be the first animals to return to native habitat in Osage County.

Because they are breeding stock, these bison are better off than they look to my amateur eyes. Each has been carefully chosen for bloodlines and health. Transponders holding microchips have been inserted under their skins, giving every animal a unique alphanumeric code corresponding to a coded computer file that includes breeding information, age, and origin and place of the mother herd.

Although soon the three hundred will roam the preserve's prairies and creek bottoms in relative freedom, they are much touched by humans and thus not quite wild. The bison we humans control—which means *all bison*—stand between the true wild animals of the past and beasts bred for human uses such as horses, goats, cows, and sheep. Even the free-ranging bison of Yellowstone Park are harassed by tourists and may be killed by authorized hunters when they forage on private or BLM grazing lands outside the park's invisible boundaries.

Having observed animals in the wild, I cringe in zoos, look away from elk farms, feel disgust at bear parks and roadside snake pits. But for creatures on the verge of extinction, zoos may be the only breeding ground for future generations. If it had not been for preservationists in the early 1900s, such as Martin Garretson of the Bronx Zoo, and western herdsmen such as Buffalo Jones from Garden City, Kansas, or South Dakota's Scotty Philip, there would be no American bison. Not one. Michel Pablo of Montana's Flathead tribe saved their herd and sold six hundred head to the Canadian government to stock Wood Buffalo Park—the world's largest bison preserve. Texas cattle baron Charles Goodnight began penning wild bison in the 1880s and developed his herd until his death in 1929. Goodnight's breed line helped stock Yellowstone Park's bison and many more.

There are nearly a quarter million bison in the United States today. Indian tribes, small entrepreneurs, and celebrity ranchers such as Ted Turner raise bison for meat and hides, as historical curiosities, for cultural reasons, for pleasure, and to entertain a tourist industry that prizes Americana. In national, state, and nonprofit preserves such as the Wichita Mountains Wildlife Refuge in Oklahoma, or Wind Cave National Park in South Dakota, or The Nature Conservancy's preserves in the Dakotas, Nebraska, Kansas, and Oklahoma, bison are bred for their own sake, and for scientific, educational, and ecological purposes. On grasslands and in mountain valleys that were part of their original range, bison again roam with deer and antelope, wolves and coyotes. There, we like to believe, they are almost free.

All this was foretold in prophecies from Plains tribes. The story I know best was told to Frank Linderman by Plenty Coups, a revered chief of the Crow Nation. When he was a boy Plenty Coups went into the Crazy Mountains of Montana on a vision quest. He fasted for four days and nights, yet no animal spirit spoke to him in dreams. Then he cut off the tip of his little finger, shook his blood on the ground, fasted, and slept again.

The spirit who came to him in a dream that night was Buffalo—an animal who'd transformed himself into a man wearing a buffalo robe. Buffalo led Plenty Coups into a dark cave.

> *I could see countless buffalo, see their sharp horns thick as the grass grows. I could smell their bodies and hear them snorting, ahead and on both sides of me. Their eyes, without number were like little fires in the darkness of the hole in the ground.*

Plenty Coups followed the Buffalo-man through the herd until he emerged into sunlight at a place called Castle Rock by white men, the fasting place by Crow. The Buffalo-man shook a red rattle, sang a song four times, and pointed back into the cavern.

> *Out of the hole in the ground came the buffalo, bulls and cows and calves without number. They spread wide and blackened the plains. . . . When at last they ceased coming out of the hole in the ground, all were gone, all!*

The Buffalo-man shook his rattle again and said, "Look!"

> *Out of the hole in the ground came bulls and cows and calves past counting. These, like the others, scattered and spread on the plains. But they stopped in small bands and began to eat the grass. Many lay down, not as a buffalo does but differently, and many were spotted. Hardly any two were alike in color or size. And the bulls bellowed differently too, not deep and far-sounding like the bulls of the buffalo but sharper and yet weaker in my ears. Their tails were different, longer, and nearly brushed the ground.*

The Buffalo-man shook his rattle a third time. Plenty Coups "saw all the Spotted-buffalo go back into the hole in the ground, until there was nothing except a few antelope anywhere in sight."

"Do you understand this which I have shown you, Plenty Coups?" asked the Buffalo-man.

"No!'" I answered. "How could he expect me to understand such a thing when I was not yet ten years old?"

Plenty Coups would grow up to see the last herds of wild bison replaced with spotted cattle. He led his people when they were independent, and was confined with them on a reservation. Plenty Coups saw the earth change under his moccasins, but although he became very old, he did not live to see his people emerge from the hole of their adversity, or witness the return of the buffalo.

Near the end of the nineteenth century, when Indians clustered hungry, diseased, and bent with loss on sad reservations, a spiritual revival swept from Nevada to the Dakotas carrying one last surge of hope. The Messiah was Jack Wilson, known as Wovoca, a visionary Paiute from Nevada. He prophesized an imminent transformation of the earth that would sweep away white invaders and return the land to its Native inhabitants. The dead would return in glory, bringing with them the plentitude of a natural world that had also died—elk and deer, grasslands swarming with buffalo.

The ritual necessary to incite this cataclysmic renewal was the Ghost Dance. Although tribes added variations in dress, songs, and preparations, the form remained uniform: Dancers painted themselves with sacred red clay; they dressed in robes edged with feathers and adorned with symbols of the sun, moon, and morning star, as well as Turtle who symbolized earth, Messenger Crow whose black wings connected the ghost world with the living world, and Eagle—symbol of the Great Spirit.

Men, women, old people, and children joined hands in a circle. They danced four nights and a fifth morning. Seven priests with eagle feathers hypnotized susceptible dancers, who fell into trances, dropped to earth in postures of death, experienced visions, then rose to sing songs reporting what they had seen and heard in the ancestral hemisphere. On the fifth day everyone bathed to cleanse whatever evil might still adhere to their bodies. Devotees performed this ceremony for six weeks.

In Oklahoma Territory the Ghost Dance was a last resort for disenfranchised Comanche, Arapaho, Cheyenne, Pawnee, Kiowa, and Caddo. The more wealthy and landed, less needy Osage held one dance cycle in the Big Hill country on Sycamore Creek, then skeptically gave up the practice. Meanwhile, starving and desperate Sioux in the Black Hills came to believe their Ghost Dance shirts were bulletproof. The Sioux Uprising, which began in dancing and faith, ended in the massacre at Wounded Knee.

The Ghost Dance fervor disintegrated. Dead Indians stayed dead. Tallgrass prairies became the agribusiness heartland of a nation devoted to cities and industrial development. Of an estimated thirty to sixty million buffalo, six hundred survived. Now we are dancing them back.

I return to the Tallgrass Prairie Preserve on a frosty fall afternoon in 1995. More than six years have passed since The Nature Conservancy purchased the Barnard Ranch and the place has remained a profitable cattle operation, but it has also become a breeding ground for bison, scientific research center, tourist facility, and experimental grasslands laboratory.

The core staff and crew are a capable bunch—stalwart and humorous. Perry Collins, the assistant foreman, lives with his wife, Danna, and their children in a white frame house adjoining the preserve's Headquarters. There is a swing set in the yard, Halloween pumpkins at the doorway, and a stuffed scarecrow. To my surprise, I also see a couple of dozen bison clustered along the reinforced fence as if waiting to be fed.

"What is this—a zoo?" I ask Perry, who has driven up in one of the white Conservancy rigs recently donated by General Motors.

"Just moved them into the new bison pasture," says Perry, a stocky, affable man wearing Carhartt coveralls and a cap. "Guess they're curious."

These free-ranging bison are wonderfully sleek and healthy; their ruddy winter coats shine as if just groomed. It's hard to believe they're the same poor animals I saw pacing their pen a couple of years ago. Many youngsters are poking alongside their moms.

"They did real good, this year," Perry says. "There's 120 summer calves. Lots more than we expected.

"We're going to round 'em all up next week. Cull about sixty bull calves and yearling bulls. Keep the heifers." Perry grins wide. "It's a new experience every day."

I wish I could be there to witness the roundup. It will take place in the new bison trap and the strong pipe corrals and pens that the crew has been building all year. When you see that setup looming huge on the plains, you realize the difference between herding cows and herding bison. The holding pens are equipped with thick metal shields, scratched and gouged by the bison's deadly horns. The head-high pipe fences and gates look more like a fortified prison than any corral you have ever seen.

Perry and his men will mount their four-wheel all-terrain vehicles and pickups (horseback bison herding being a rarely practiced, dangerous, and lost art) and push the 450 beasts into the enclosure. Sometimes, Perry explains, you have to stand up in your ATV to see over the quick-footed, huge humped animals. It can be more thrilling than anyone hopes, because even on wheels a man is no match for an enraged bull.

"Rounding up bison with a four-wheeler," adds the foreman, Kenny Shieldnight, "you have to keep them front to you or back to you." I gather you don't want to be sideways to those horns.

Kenny is a blue-eyed Oklahoman in blue denim, his round face weathered in smile lines and crinkles—a cowboy conservationist. He and his wife, Royce, and their daughter Billie Jean live in a rambling ranch house near the entrance to the preserve. A dayroom lined with saddles and tack leads in from the driveway and barns. There is a long Formica kitchen counter, and beyond it an open living room carpeted in shag and furnished with plaid-covered easy chairs.

Royce is lean and narrow hipped, with shoulder-length gray-streaked hair. Raised on New Mexico ranches by a cowboying family, she does the cleaning chores at the preserve's Headquarters, cooks hearty lunches for the hands, and helps out with the horses and cattle that she and Kenny raise on their place. Royce exhales a small cloud of smoke. "I love the space and the freedom out here."

Kenny goes to the kitchen to fetch a cold, end-of-the-day beer. He tells me the work on the preserve is not that different from work he's done on other big ranches: building fence, running stock, building more fence, but this time for bison. "It's still a ranch," he says, "and I'm the working end of it.

"Cattle business, you've got a lot of burnout in it. Cattle day after day after day." Here, Kenny is not bored. "When I was cowboying," he continues, "there was grass, and there was weeds. Now I know all the names of the grasses, and the weeds have become wildflowers."

At first Kenny and his crew were hassled by other ranchers and cowboy neighbors—mostly because the preserve didn't run any cattle for a couple of years. "Some guy'd come along and ask what I was doing." Kenny laughs. "I'd look him in the eye and say, 'I'm countin' butterflies today.'

"I'm not retired," he says, grabbing up his daughter, "I'm going to die out here. Only problem is me and my girl don't have enough time to go fishing, do we?" The child shakes her head, curls flying. "The truth is, I believe in what we're doing."

Although I am still a stranger in these parts, I have a powerful sense of coming home as I hike up into the hills this sere November evening. The woods open when I reach the ridge and a soft rain mists the prairies. I see the deep sienna color of oak leaves and the orange-tipped bluestem grasses. Thunder booms in the distance, and lightning is a far-flung flash of blue.

Below me in the valley are the yellow lights in Perry's house and the shuttered windows of cowboy shacks behind it. I see dappled and roan horses grazing in the pasture. Coyotes sing to each other and a great horned owl rests stone-still in the arms of a lone post oak. A deer crosses my path and disappears into the undergrowth. Although there is barely a breeze in the rain-thick air, when I walk between twin oaks that arch over my head like a tunnel their leaves rattle, hiss, and whisper as though inhabited by ghosts.

Next morning is Sunday, overcast and drizzling. I head for the crossroads where bison are grazing last summer's burn. Pickups pass me on the gravel road, drivers waving in the habit of the country. A white Buick full of white-haired oldsters on a Sunday drive cruises slowly by. The cars pull to a stop near a patch of blackened ground where sixty bison quietly graze the green new shoots that have emerged in fire's wake.

I join the folks who stand almost reverently before the bison. Some approach them with cameras. Some stay back, content to see the humps, horns, and wide-set eyes through binocular lenses. A few hunters in camo gear and scarlet vests pile out of a pickup. Everyone whispers as though in church. Parents with small children hold them close.

From distant ridges, I hear the *chuck-chuck-chuck* of oil pumps beating like the Osage drums of another century. I realize I will never see what the Osage saw, or know what the prairie was before humans stepped into it, but I'm happy to be standing where I stand. This is the prairie of now—a compromise that seems to be working. City people are here, and local folks, foreign tourists, and hunters who have come to see the native beast on its native land, never mind roads and pumps and telephone wires.

Later, near the abandoned foundations of the ghost oil-boom town of Pearsonia, I take my camera out, for I have happened on the perfect symbolic image. But I'm out of film. I decide to chase the image anyway, imprint it on memory, where it belongs.

I walk down a draw through big bluestem stalks seven feet high. Tallgrass drips from this afternoon's rain, and my jeans are soaked. But on the western horizon black clouds are rising up the sky, dispersing.

Across a wetland creek bottom stands a bull bison, his thick brown winter coat fringed in the tones of a lion's mane, backlit by the setting sun. He will be my foreground. Behind him is an abandoned oil field with rusting pumps arrested in midmotion like huge pecking crows. Still farther back, in middle perspective, is the wide prairie speckled with grazing cattle. But it is the far distance that astonishes. Here is an apricot sky running with silver-edged clouds, washed with patches of bird's-egg blue. The heavens are opening. Under wings of light, grasslands spread in the four sacred directions.

I stand in the tallgrass after rain, feeling as insignificant as a grasshopper or a spider under the sun, happy to be enveloped between the great earth and great sky. This evening, red light seems to be emerging from the prairies. Bison graze in glowing grass. For a moment, the world preens in absolute beauty.Then it dims. Darkness follows sunlight as ash follows flame.

III. Seeding with Fire

And the flames folded, roaring fierce within the pitchy night . . .
In flames as of a furnace on the land from North to South.

William Blake, from "America"

A pall of smoke hovers over Pawhuska town. As I pass through blackjack thickets on the skirting hills, I smell the grasslands before I see them. Then I'm driving amid clouds of brown vapors edged scarlet with tines of flame. It is late March, and the flames are burning winter's tallgrass.

This is the season of fire and birth. After snow and ice melt, while new grass is germinating, the old grasses lie in a tinder-dry mat upon the earth. With no wet green growth to hinder its spread, a March wildfire lit by lightning sweeps across the prairies until the flame front is doused by rain or hits a moist riverbank, a swamp, a lake, or rocky ground. A grass fire in early spring is like Blake's fiery furnace, running with the wind.

The fires I pass through today are not wild, but human caused and controlled. Every rancher in Osage county seems to be firing his pastures. Even on the tallgrass preserve, cattle grazing areas are aflame, for spring fires spur the Big Four prairie grasses to be reborn in profusion. Big bluestem, especially, responds well to fire.

"The tallgrass prairie has always been a disturbance-dependent ecosystem," Bob Hamilton explains. "Fire is the most dramatic recurrent natural disturbance, intensive grazing is the other." In the old days bison herds followed fires; where bison did not forage, grasses dried in deep layers, creating fuel for future fires. The chain went on and on, and the prairies stayed healthy and diverse. When the bison were killed off, cattle became the next best grazers. Cattlemen have been firing the Osage since the Reservation was allotted a century ago.

"I've set a lot along this ranch," says cowboy Dink Talley. "Old man Chapman used to burn it off every year. He'd make fire balls out of gunny sacks. He'd fold 'em up so they wouldn't be much bigger than your hand. And soak 'em in kerosene and motor oil, crude oil and everything else. [And I'd] fasten one up on a rope, and start out with it. I drove one up five miles before it burned out, one time, down the creek."

In ranching setting spring fires is commonplace, but no one wants an uncontrolled burn—especially in summer, when grazing is prime. Wildfires wipe out homes and barns, kill horses and cattle, and endanger families. As Americans colonized the West, building communities in wild grasslands, forests, and chaparral, fear of wildfires led to fire suppression, which led to a catastrophic buildup of fuels—most direly in our national forests.

After a century deprived of fire, nature is taking revenge. We who watched Yellowstone burn in 1988, or were smothered in the smoke of the millennial Bitterroot blazes, know the terrors. Heat becomes so fierce it sterilizes soils. Crown fires running in treetops devour not only weak but also mature trees. Heat and smoke choke animals, insects, and plants. After high-intensity forest fires, nature takes a long time reconstructing a system of diversity.

When forest fires are frequent, fuels do not accumulate and intensity is lower. Sporadic, low-intensity burns enrich soils with ash, clear brush, promote grass, toughen healthy trees, and do not permanently diminish animal or insect life. Biological and plant diversity are enhanced by new growth in climax forests. We see redistribution of insects, rodents, birds, mammals; patches of pioneer plants; and savanna openings in deep woods.

Fires in grasslands run with different rules. Even if fuels are heavy and heat is intense, the sod acts as an insulator. Heat during grassfires does not usually get high enough for a long enough time to sterilize soils. When I walked into a meadow a couple of weeks after it was fired, pale leaves of grass were sprouting through the ash, like lettuce in my garden.

"Hair," the ranchers call this grass, as in, "The south pasture's haired up enough to move the cattle in." I like the notion of grass as hair. Cut it, and it grows back. Burn it, and it grows back. That's because the heart of grass lies in its roots—with tallgrass, really deep roots—and the growing points lie under the sod that a root system builds.

Trees are different. Even hardy, deep-rooted bur oaks expose their growing points and buds. Drought and fire will kill trees from the top down, especially in summer and fall, when they're in full leaf. Although necessary for the germination of species such as lodgpole pine, fire kills an oak tree's seedlings. It brings down forests and opens grasslands to sun and rain, promoting more grass. Which is why, if you came to the preserve as I did on a mild and windless day in summer, you might witness a wall of flames. Don't worry. Though the fire's wind roars in your ears; though its smoke sets you to choking; though its color is scarlet and clouds of sulfurous fumes burn to beat the sun, this fire has been mapped and choreographed as elegantly as a ballet.

Bob Hamilton has set this controlled burn on a piece of ground bordered by exposed sandstone hills where cross-timbered blackjack and post oaks grow in thickets. Beyond the tree line, brush has been creeping into the tallgrass. If nothing is done to disturb the land, the sumac, wild plum, and Osage orange will advance, creating shade for seedling oaks and transforming the sun-dried prairie to woods. Bob wants to know if summer fires will stop the brush and tip the balance of regeneration toward grass. He wants to know how many fires it will take to make a difference.

The grass where I stand is so tall I cannot see over it. I could be engulfed by flames with no sense of where to run for safe ground. Bob assures me I'm safely outside the fire lines but I am not reassured, so I scurry toward a rocky mound. In the near distance, scarlet wings are flying twenty feet high. White smoke billows a hundred feet into the air, melding with black smoke, darkening the sun-bright day. The blazing wave moves fast. When the front hits a gully, the inferno roars with the wind it has created. Gully fires are especially dangerous, but Bob and his crew have three spray rigs ready to douse runaway flames, and they keep in moment-to-moment contact using hand-held radios.

I hold a wet cloth over my nose and mouth, watching grass, forbs, saplings, brush turn to ash. A few small oak trees flare. Birds rise in alarm. Mice and small rodents skitter for shelter. Grasshoppers fry. Snakes are burned into black coils.

Large animals such as deer can run from fires, as can birds on the fly. But species such as the northern short-tailed shrew, prairie voles, and tawny cotton rats, which depend for food and nesting sites on the debris of prairie floors, suffer immediately after a fire. And rodents that nest aboveground rather than deep within it, such as white-throated rats, western harvest mice, and some vole species, are doubly vulnerable. Tiny, slow mammals like these are called fire-negative.

Quick-moving rodents that prefer open grassy habitats and feed on seeds and insects thrive after fires. Included in this category are the hispid pocket mouse, southern grasshopper mouse, deer mouse, and thirteen-lined ground squirrel. Jumpers and hoppers such as Merriam's kangaroo rat and the meadow jumping mouse also respond well to fire. They're called fire-positive.

Being fire-positive does not ensure an individual's life, but it does ensure group survival and change. After fire, population densities will vary and communities will move. Fire also alters the levels and timing of reproduction. But in the long run—say, three to five years—if fire does not recur in the same place, the conditions of both plant and animal life that existed before the grass went up in flames will resume. In the meantime, other patches will have been colonized and new patterns set. The dance of prairie life circles around fire, which redesigns the habitat's mosaic and enhances diversity.

I wander away from the fired earth onto undulating meadows that roll eastward toward a rim of trees. Here a few widely separated ancient post oaks—maybe three hundred years old—stand huge in the grass. They are twisted by sun and wind. Heavy limbed. Wrinkled with age, blackened and hardened by recurrent fires in their youth. Crows gather in the high, open-armed branches above my head. I listen. I think they are crowing *savanna.*

Savanna is one of those connotative words that sing like the partially remembered music of childhood. The word conjures baobab trees under a red sky. Beneath the trees are a sweep of tall grasses, yellow and bent by wind. And hidden among them are honey-maned lions with their mewling, playful cubs. I see elephants, gazelles, giraffes.

Africa, first Eden, is not so different from the red-fringed Oklahoma prairies that surround me. In savannas, at the edge of trees, looking out to hills alive with gilded grasses, the upright ancestors of human beings emerged from dark, smoky forests. I can imagine them here, squinting into bright sunlight, carrying spears in one hand, fire in the other.

The Osage myth of creation says the Little Ones came down from the stars, but in the tale of origins Science tells—at least in the one I am about to relate—the great force behind human evolution was Fire. In an intriguing essay called "Landscapes and Climate in Prehistory: Interactions of Wildlife, Man, and Fire," W. Schule of the Institute of Prehistory at the University of Freiburg offers a compelling theory about the intertwined evolutions of landscapes and life. Schule says:

> *Landscapes are more than a simple function of geological, geomorphological, climatic, and botanical parameters. Animals play an important role. Their behavior, especially their trophic habits, is a major force in the forming of landscapes. Herbivores consume the product of the primary biomass production. Fire and man have been doing the same since they appeared on Earth. Moreover, both are not only herbivorous, but carnivorous, devouring whatever animal wherever they can.*

All creatures except our kind are so afraid of fire they will flee if they can. We are afraid too, but also enticed, for we have learned to use fire as a tool. Early upright walkers developed a taste for the fresh-singed remains of huge animals. They alone among primates discovered the pleasures of roasted prey. Fire was also used in hunting. Hominids could not outrace swift animals, but they could trap them in encircling flames. And fire was a safety device, keeping wolves, lions, cave bears away from communal shelters. It is no accident that the symbol of family and home is the hearth.

Dr. Schule and other scholars believe that some of the earliest human settlers on the American continents were upper Paleolithic hunters who had emigrated from Eurasia. According to this theory, about twenty-five thousand years ago they occupied Beringia—an ice-free arid grassland that stretched from Siberia to Alaska. Eventually, following prey, Paleolithic clans crossed over the Bering land bridge to North America. Scientists have named these hunters Clovis People and identify their remains by a genetically distinct shovel-shaped hollow teeth.

Recent discoveries have unearthed remains of others in the Americas who predate or coexist with Clovis People, and while we hope to learn more through DNA and carbon dating, no one is likely to discover exactly who inhabited prehistoric America or when they came. Nevertheless, we do know that colonies of Clovis Paleo-Indians were confined on the Arctic steppes and North Pacific coasts by the great American ice sheet. Then, probably about twelve thousand years ago, the ice melted enough to open a passage west of the Rocky Mountains, and the hunters wandered south.

The lands they found were temperate, with open forests and grassland savannas, like Africa. But unlike Africa, Europe, and Asia, where Neanderthal and other hunters had existed alongside their prey for eons, this was virgin territory. Here were gigantic herbivores and smaller ungulates that had never (or rarely) been hunted by two-legged creatures and had no inherent fear of them. What a paradise for America's Adams and Eves!

But I am oversimplifying. It was no easy task to puncture mammoth hide with handheld spears. Courage, strength, cunning, and perseverance were needed to kill the roaring beast. We know that Clovis hunters were often successful. Digs in Nevada and throughout the continent have unearthed mammoth bones chipped by obsidian Clovis points. Probably they were too successful.

Fossils indicate that during three million years, right up to the end of the Pleistocene, only about twenty genera of large mammals became extinct in North America. Then came Clovis hunters, who were skilled in the use of spears and fire. Many paleontologists believe that within a few thousand years they exterminated, or were influential in exterminating, all the giant mammals in the Americas, several species of huge flying and flightless birds, and some of the continent's largest reptiles. Thirty-three genera died out, including about one hundred species. There would be no more American horses, camels, *Bison antiquus,* mammoths, rhinos, dire wolves, saber-toothed tigers. The only large native grazing mammal to survive was the antelope, who could outrun hunters and fires.

Some experts believe climate may have played the strongest role in the animal holocaust of the Americas, but paleontologists such as W. Schule, Paul Martin, and E. O. Wilson lay the blame mostly on humans. Whether or not we're totally to blame, heedlessness seems our oldest sin. Perhaps we were always voracious, never innocent.

Then times got tough. With few animals to control the biomass, fuels built up. Changing climate brought drought, and fire intensity in forests and grasslands became devastating. Vegetation was deleted along with animal life. Starvation ruled for hunters and beasts. But not for long.

With less pressure from herbivores and human predations, the North American forests and grasslands recovered. Now, however, the woods were dense, the prairies invaded by brush and small trees. Eurasian species such as bison, elk, bear, and beaver came down from the north. Again came northern hunters tracking their prey. But this time the bison, deer, bear, and elk were also immigrants. They were smaller boned then the original American herbivores, swifter, wary of humans, and fire-wise.

Bison, especially, thrived in the North American heartlands. Intensive grazing by herds of bison in the millions jump-started grass production and changed the landscape. Their grazing habits lowered accumulation of fuels and diminished the intensity of fires, if not their frequency, becoming a major factor in the shaping of a prairie ecosystem.

The American prairies may be imagined as vast grass and flower gardens sown by fire. Gardens need cultivation, and the hunters, who were also gatherers, became domesticators and agriculturists. One of their primary instruments was the burning tool. Like earlier Clovis hunters, Plains tribes and immigrant tribes from east of the Mississippi used fire for hunting, war, self-defense, and to increase grass. At times the fires turned destructive, but during their primacy on the prairies, American Indians learned to control the force they called Red Buffalo more wisely than their Clovis precursors. Preservation of resources meant preservation of lives.

Indian fires helped clear the grasslands of trees and shrubs, and thereby helped feed bison and elk—the tribes' best sources of protein. Even when the climate became cooler and wetter a couple of thousand years ago, promoting a return of trees, Indian fires were instrumental in holding the prairie system to its old drought-induced limits.

When Spanish and French explorers and traders ventured into the prairies and woodlands west of the Mississippi, they noted the smoke of Indian fires. Joliet mentioned wildfire, but did not comprehend its importance. Lewis and Clark were more aware. Their journal entry for March 6, 1805, reads: "a cloudy morning & smoky all Day from the burning of the plains, which was set on fire by the Minetarries for an early crop of grass, as an inducement for the Buffalow to feed on. . . ." In a letter to his mother, Lewis elaborated the theory that western prairies were formed by Indian fires.

Scientists are still debating the role of Indian fires in creating prairie landscapes, but it is obvious that Indian fires would have had little effect without the overriding influence of drought. Sun is the force responsible for drought. Sun rules the prairie. We must thank our higher powers that no person controls the sun.

I will tell you a cautionary tale about the Sun and his children. It is not my story, but Aesop's. In Greece, about six hundred years before Christ, Aesop was a Phrygian slave. His tales are ancient teaching narratives about natural forces.

Once upon a time the Sun was about to take to himself a wife. The Frogs in terror all raised their voices to the skies, and Jupiter, disturbed by the noise, asked them what they were croaking about. They replied, "The Sun is bad enough even while he is single, drying up our marshes with his heat as he does. But what will become of us if he marries and begets other Suns?"

Drought, lightning, and wind may be understood as the offspring of Sun—forces that created grasslands and thus, in a way, created us. Corn, on the other hand, is largely a human creation, a domesticated tallgrass that has become the supreme crop on the breadbasket prairies of the Midwest. But unlike drought-resistant big bluestem, corn needs constant moisture. The rich midwestern plains that are planted in corn are prone to drought. Ask any farmer about drought. He will look at the sky and throw his hands out, palms up in a who-knows gesture. Even if drought means famine, what can he do?

Lightning is an equally powerful climatic commonplace on the plains. The Osage prairies of Oklahoma and the Kansas Flint Hills to the north are lightning country. In 1989, according to the U.S. Weather Service, this region drew the second highest number of lightning strikes in the United States. Add the elements of drought and lightning to grass and you've got a recipe for fire.

"Every light against the sky told of a prairie fire," wrote one Kansas pioneer woman. "The direction of the wind, either from or opposite the direction of such fire, or sidewise, the unsteadiness of wind with possibility of veering so as to bring fire toward the home—all these were noted. . . . Many times, on awakening in the dead of night, the room was light with reflection from the sky, shining thru uncurtained windows from some fire ten or twenty or fifty miles away."

Wind, ever blowing, it is said, drove prairie women crazy. Wind in the extreme, together with electric storms and rain, means tornadoes. Weather-crossed Oklahoma—along with Missouri, Illinois, Kansas, and parts of Texas—is known as Tornado Alley. Since 1950 the average number of twisters to hit down on Oklahoma has been fifty-three per year. Many more people die from tornadoes than from prairie fires. The most deadly swarm of tornadoes recorded in the United States swept through prairie country from Missouri into southern Illinois in 1924, killing 689 men, women, and children.

In John Madson's *Where the Sky Began,* there is an account of a Kansas farmer—Will Keller from Greensburg—who claimed to have looked up into a cyclone's eye. Keller reported:

> *. . . At last the great shaggy end of the funnel hung directly overhead. Everything was as still as death. There was a strong grassy odor and it seemed that I could not breathe. There was a screaming hissing sound coming directly from the end of the funnel. I looked up and to my astonishment I saw right up into the center of the funnel, about 50 to 100 feet in diameter, and extending straight upward for a distance of at least one-half mile, as best I could judge under the cir-*

cumstances. The walls of this opening were of rotating clouds and the whole was made brilliantly visible by constant flashes of lightning which zigzagged from side to side.

Tornadoes, lightning, and drought are forces of nature, and destructive, but they do not hold a candle to the environmental damage we humans have done, and continue to do. People created the conditions for the 1930s Dust Bowl disaster in the wheatlands of Oklahoma. Plows, not fires, destroyed deep-rooted grasses; plows ripped open the prairie's stabilizing sod skin, baring soils to constant drying currents. Whirlwinds of eroded topsoils that brought darkness at noon and drove thousands off the land were occasioned by human hubris, error, and greed. But plows and the people who use them did not create the wind itself or the years of drought that dried soil to fine powder.

Today, earth is warming from the carbon-based energy we've used in industry and development. We've made holes in the ozone and tampered with the natural world to such a degree that we can't stop controlling it, but it is reassuring to know that our species is not in charge of the wind, rain, or sun. For no matter how lovely our rain dances, and despite science and technology, when cosmic forces take over, all we Homo sapiens can do is croak for help to Jupiter like Aesop's lowly Frogs.

I'm glad groups such as The Nature Conservancy are working to restore prairies, protect grizzlies, plant forests. This is human control that enhances life, maintaining the biological diversity necessary for survival on earth as we know it. But life will go on without tall-grass, without bison, without us. The sun doesn't care. In the sun's time the voracious millions of all humankind will be mere specks in one layer of stratified rock. Humility is the lesson fire teaches. The universe is made of fire.

PART 5

PABLO NERUDA'S HOUSE IN THE SAND

... Perhaps it was only a long day the color of honey and blue ...

Pablo Neruda, from "Love for This Book"

White-shirted peddlers enticed us with fresh, ripe strawberries. Boys held out baskets of warm empanadas or coolers filled with ice cream, but we kept our car windows up, the air conditioner on. Weekend tourists crowded the roads as we passed through sandblown towns (San Sebastian, Las Cruces, El Tabo) with their shuttered cottages, roadside eateries, and wide public beaches. It was a blazing summer Sunday in Chile, mid-January, and Bill and I had driven our rented white Suzuki some sixty miles on mostly good four-lanes from Santiago to Cartagena. Then, funneling onto the narrow, jammed, coastal road, we had only a scant ten miles to go.

We were heading for Isla Negra, the seaside home of the poet Pablo Neruda, now a museum. Neruda had built his sprawling estate on a wooded peninsula set apart from the peopled lowlands and overlooking the wildest seascape to be found on that coast. I'd been a longtime fan of the poet's works, and was reading his memoirs, studying his life, hoping to better understand both the poet and the place that had formed him—a Pacific shore like California's, yet distant in every particular from politics and history to flora and fauna—so I was eager as any pilgrim to visit the dwelling place of the master.

When Neruda bought the original stone house and three acres at Isla Negra from a retired Spanish sea captain, he was thirty-five and already a well-known poet in South America, but just beginning to reap substantial royalties. The year was 1939. He had returned to Chile from Spain, where he'd been embroiled in the Civil War, opposing General Franco and the Fascists. After his friend Federico García Lorca was assassinated in Grenada, and after witnessing death in besieged Madrid, where "in the streets the blood of children/flowed simply, like blood from children," Neruda became bitter and angry. His impassioned collection *España en el corazón* marks the transformation of a personal poet into a public poet—a man determined to act on behalf of the poor and the powerless.

Neruda joined the Chilean Communist Party around that time, an act that reminded me of my father, who'd become a Communist in Paris in the 1930s, wanted to go fight Franco in Spain, but was convinced by my mother to emigrate to Chicago instead. My father stayed Red until the party kicked him and his renegade Chicago comrades (Nelson Algren, Studs Terkel, Ben Burns) out for refusing the Stalinist line. Neruda, however, remained a faithful member for the rest of his life and was elected to the Chilean Senate representing miners in the northern high deserts.

During his long political tenure he also served as consul in nations as various as Mexico and France, and in 1970 he reluctantly agreed to run for president, then dropped out of the race to join forces with a leftist coalition whose Socialist candidate, Doctor Salvador Allende Cossens, was his good friend. Allende won the election and set about nationalizing the industries in Chile, while in 1971 Neruda was awarded the greatest prize a poet can win—the Nobel Prize in Literature.

Neruda's life was as adventurous as a fictional hero's and romantic as a movie star's, yet his character remained down to earth. "As for me," he wrote, "I am—or think I am—hard-nosed, small-eyed, sparse of hair, swollen in the abdomen . . . generous in love, impossible at figures, confused by words . . . fond of stars and tides and swells, an admirer of beetles, a walker of sands, institutionally dull, perpetually Chilean. . . ."

When he was in his sixties Neruda retreated to Isla Negra, where he could live and write in relative solitude. Looking out a study window at the waves constantly born and reborn, he thought about death and the place where life was good and wrote *The House in the Sand,* a book of prose poems. On his desk was the cast of a woman's hand—that of wild-haired Mathilde, his third wife and final love. Her hand helped him to write, he said, and kept him company when she was away.

Neruda wrote that Isla Negra had grown "like people, like the trees," book by book, as his fortunes increased. It took shape like an epic, mirroring his history, his fantasies, the images and metaphors that appear throughout his rich body of work. To wander amid the rooms of Neruda's three houses (Isla Negra by the sea, La Chascona in Santiago's Bella Vista district, and La Sebastiana on a Valparaiso hilltop) is akin to wandering in the baroque labyrinths of the artist's mind. Neruda was first an architect of words, second a revolutionary who wanted to rebuild societies, and third a maker of homes.

He exalted the crafts of building, writing poems to Alejandro García, the stonemason who built the tower at Isla Negra, "hefting the paving block, cutting the granite grapes," and to Rafita Plaza, "poet of carpentry," who is still working in the neighborhood. He also wrote about objects he had discovered and installed in his house: a carved Elizabethan figurehead buried in sand by mutinous sailors because they believed her beauty had steered them off course; a giant wooden key from his childhood town of Temuco; the whale's tooth inscribed by a sailor with images of his betrothed, which once sat on his table "facing the waters of March"; and a wooden Indian he had purchased in Boston, which he imagined had been carved for Nantucket whalers.

These artifacts inhabited his rooms, as did the spirits of dead invisible friends—García Lorca, Nazim Hizmet, Paul Eluard, Miguel Hernández, Joaquin Cifuentes, Rojas

Giménez—fellow poets whose names Neruda carved into the rafters so they would keep him company as he drank the red wine of his country, served warm, as he liked it.

Birds, wildflowers, shells, and the agates on the beach where he strolled each morning and evening were also subjects for contemplation, uniquely *Chileño* and strong in his imagination. Nothing, however, could take precedence over the sea. Again and again Neruda tried to capture its essence. (He prized everything about the sea except immersion in it—a fear derived from being dunked in cold water by his father.) "The sea sings and pounds," he wrote, "it is not in its right mind. Don't tie it down. Don't lock it up. It is still being born." At night the sea stole into his room and coated every object with its salty, teary breath. *The salt of seven leagues, horizontal salt, crystalline salt of the rectangle, stormy salt, the salt of the seven seas, salt.*

Pablo Neruda's only child had been a daughter born encephalitic, who died when she was eight. Enrique Salazar, later the caretaker at Isla Negra, was a neighbor boy who became a kind of foster son to Mathilde and Pablo for twelve years. One of his jobs was to help care for the dogs. Neruda always kept dogs, longhaired chows who chased after sheep until a fence was built, not to keep people out, but to keep the dogs in. He wrote about "this gateway toward the limitless"—not knowing the fence would become a famous message board, a kind of shrine.

On September 11, 1973, an army coup tacitly supported by the United States (via the CIA) to protect its multinational corporations in Chile (such as the Anaconda Company) overthrew the democratically elected president, Allende. He was shot and killed in the seat of government in La Moneda, some say a suicide, but Neruda and many others believed the president had been gunned down by the army. General Augusto Pinochet took over the government, and under his reign opponents were hunted and killed. Thousands "disappeared."

At the time of the coup Neruda was deathly ill with prostate cancer. He lay sick in his bed at Isla Negra when Pinochet's soldiers came to search his house. They examined his first editions of Whitman and Rimbaud, peered into seashells, poked around the steam locomotive in his courtyard looking for contraband weapons. In a legendary story about this encounter the commander tells Neruda that informants have reported he is hiding something dangerous, perhaps guerrilla Communists. "Yes," Neruda replies, "very dangerous indeed. It's called poetry."

Soon afterward, Neruda's fever rose. His physician had been arrested and could not attend him at Isla Negra, so, refusing invitations to fly to a hospital in Mexico, where he would be safe, Neruda set off in an ambulance for Santiago. En route the ambulance was searched twice by soldiers and was delayed three hours. Mathilde said it was the first time she'd seen her husband weep. A few days later, on September 23, Neruda died. Friends say he died of a broken heart. His last reported words were: "They're shooting them, they're killing them."

Although the army, the navy, and assorted thugs vandalized and nearly destroyed Neruda's La Chascona home in Santiago and La Sebastiana in Valparaiso, Isla Negra was boarded up and remained relatively intact. People could not go inside to pay their respects, but year after year during Pinochet's oppressive regime, on July 12, Neruda's birthday, in the dark days of Chilean winter when cold winds and rains blow in from the sea, visitors would come to the poet's house by busloads and carloads. They pinned paper messages to the wooden slats of his fence, carved words of love into posts, scrawled lines in charcoal that would be washed away and replaced with new messages, new prayers as the years turned toward freedom.

Twenty-six years after the coup, in a Chile now run by a democratic government, Bill and I walk toward Isla Negra. There's still a fence there, and on the day we arrive it is festooned with children's paintings. A carnival feeling surrounds the grounds. Families and teenagers, young lovers and old folks are heading down a path to the beach and we join them, although we have foolishly left our bathing suits in the car.

I take off my sandals, and the sand between my toes is coarse and distinctly yellow. According to Neruda, the particles are made of yellow granite and are unique. "The golden sands of Isla Negra," he called them, and I am happy to be walking where he walked, where laughing crowds are jumping into the surf (not swimming, for it's too rocky and rough to swim).

Pulling my green polka-dot sundress up to my

thighs, I test the water. It is bright blue, sparkling, cool but not cold. A grandmother hoists her skirts next to me. The surf crashes foam-white on the rocks, splattering both of us. Her pigtailed *nieta* laughs. Farther up the beach three brown-skinned girls in pink and lime and tangerine bikinis preen in the sun. And sprawled on a rocky peninsula, a fleshy family is finishing a picnic lunch, hooting like a pod of sunburned seals.

Bill wanders toward a huge rounded hunk of granite that is covered with painted images of what looks to me like Fidel Castro and a Mapuche Indian woman and Uncle Sam with a death's head under his red, white, and blue top hat. The rock is crowned with a massive carving of Neruda's head—the distinctive cap, beaked nose, and hooded eyes (Neruda often compared his face to that of a turkey). Two small boys have scrambled up there. The one in blue shorts sits plumb on the poet's cap. I take his picture. Neruda would have liked this carefree spectacle, I believe, everyone equal under the sun, partway naked and reveling.

We ascend the path toward Isla Negra, which is not a single house but a joining of square, rectangular, and circular buildings linked like children's hands in a staggered line on a bluff above the sea. Picture windows and French windows and curved windows and windows leaded in colored glass look out from every nook and angle across the flowering green-fringed dune to the half-moon beach, curved and yellow, where the Pacific's foam explodes against dark-toothed rocks.

The oldest structure is built of stone, "rugged sparkling, grey, pure and heavy, so that you may construct, with iron and wood, a house in the sand," wrote Neruda.

It is late afternoon when we begin our tour, and we are lucky. Besides Bill and me, there are only a young American poet and his Chilean lady love. "In summer," says guide Catherine Clyner, "it's a marathon. Sometimes fifteen hundred a day."

We are standing near the entrance of the museum, built in 1989, complete with ticket booths and café. Nearby, on a patch of gravel surrounded by native pines, stands a gaily painted steam locomotive with iron wheels and a tall smokestack. It came from the south, Neruda's homeland, and reminded him of his railroader father. But most of all Neruda loved this steam engine because, he said, "it looks like Walt Whitman," the poet whom he admired and emulated above all others.

Entering the original stone house, which ultimately became the dining area (Neruda was a wonderful chef who wouldn't allow anyone in his round kitchen when he was preparing one of his feasts), I notice that the ceiling is paneled in wood and bowed like the hull of a ship. There are nautical features in all of Neruda's houses: winding stairways, sea chests, plank floors, lookouts, decks, and the beautiful bowed ceilings. The poet loved to wear masks, dress in costumes, and enact his many fantasies, the most common one being the role of sea captain. Neruda's houses were like toy ships to him, and he could play in them all day long. He even designed a flag—a fish in an orb against an azure ground—which he raised on a flagpole to tell all comers he was in residence. The design was patented and copyrighted. We see it in the weather vane that crowns the tower.

Neruda's tower is dark, but Mathilde's bedroom wing with its lace bedspread and picture windows is an airy nest. One of the house's several studies has walls and a fireplace made of agates, onyx, quartz, coal, and lapis lazuli cemented in a mosaic that flows like a river. It was designed by his neighbor in Valparaiso, the artist Mari Martner.

Neruda's last-built and favorite study was the Covacha, a cubbyhole of vertical rough timbers with furniture of knotted wood and walls decorated with signal flags and etchings of birds. The room was designed to resemble a cabin in the southern wilds of the poet's childhood—a kind of psychic home. Nearby is another nostalgic room holding a great papier-mâché horse from Temuco. And finally we are shown the new wing, which houses the narwhal horn and shell collection Neruda bought with money from his Nobel Prize. (The third item financed by this prize was a first edition three-volume set of Diderot's 1772 *Encyclopédie,* which is housed, along with what remains of Neruda's library, in his restored Chascona house in Santiago.)

It is not surprising that Neruda would spend his earnings and winnings on ancient books and treasures from the sea. From his twenties through his sixties he had traveled across oceans and continents as diplomat, exile, and bard. He was a walker of beaches, a connoisseur of flea markets, an obsessive collector of man-made things—but also of natural objects, particularly beetles and butterflies and seashells.

Neruda's mother had died in 1904, soon after he was born, and he was raised by his father and stepmother in the frontier town of Temuco. Tagging along on the logging train his father conducted, Neruda spent much of his childhood wandering the southern forests. He developed a naturalist's passion for precise seeing and naming. Julian Huxley, who was a good friend, once claimed (perhaps a bit tongue-in-cheek) that he admired Neruda the malacologist (specialist in mollusks) nearly as much as Neruda the poet.

Most of Chile's forests have been cut down, and the old frontier around Temuca has been transformed into an agricultural country, but there is still wildness in Chile's southern hinterlands—in Patagonia and the Andes. Travelers like us can see the volcanoes, mountain lakes, and preserved bosquets of araucaria pine that inspired young Neruda. And of course rain still falls in the southland, the constant rain, which, he claimed, gave him rhythm and voice.

"Anyone who hasn't been in the Chilean forest," Neruda wrote in his memoirs, "doesn't know this planet. I have come out of that landscape, that mud, that silence, to roam, to go singing through the world."

Although Pablo Neruda named his seaside hideaway Isla Negra, the place is not an island and it is not black. Like his public name (birth name: Ricardo Eliecer Neftalí Reyes Basoalto), it is a poet's invention (*Pablo,* claimed Picasso, was chosen by his friend in homage, and *Neruda*—the name of a Czech writer—rolled nicely off the tongue).

The man's penchant for make-believe is also evident in the bars in his house, the main one opening onto a stone patio overlooking the Pacific. Here are stools and tables where Neruda could imagine himself at a waterfront café in Nice or Rangoon or Havana—where we can imagine him entertaining literary guests. I have seen the photos. Angel Rama plays guitar; Juan Rulfo smokes; Mario Vargas Llosa raises his glass in a toast.

The host served French 75s (half champagne, half cognac, with a few drops of Cointreau and orange juice) while the setting sun glinted through windows stacked with ornate glass bottles. Later Neruda might lead his guests down a lane to the inn next door to feast on *camarones pil pil*—small shrimp with a warm garlic and lemon sauce—or fresh sea bass, or cold eel (his favorite). There would be tomato salad with onion and cilantro, bread and Chilean wine, perhaps flan and coffee—a common, scrumptious meal, like the one we ate in that same inn on a glassed-in porch with a view to the sea. Doña Elena, who used to run the place, often served Neruda. Today the remodeled restaurant and hotel is called La Candela (a candle lit for the people's revolution) and is owned by Charo Cofre, a folksinger and documentary filmmaker who recorded Neruda in his final years and became a political friend.

After the tour, Bill and I sit on a bench on the flat promontory that forms the plaza around which Isla Negra's buildings are strung. Above us rises a campanile hung with six bells and shaped like half of a Hebrew star.

"It's like Big Sur," says Bill. "Reminds me of Robinson Jeffers."

"Neruda said Isla Negra was built to mimic the geography of Chile," I reply. I'm thinking about how similar to and yet how utterly different this blufftop home is from my family's summerhouse on a dune above Lake Michigan. The difference lies in the scale of an artist's imagination, I think, separating the great from the good. Neruda was daring in every move. Sure of his instincts. My father, also an artist, was more inhibited in his ambition and vision, divided in his heart.

Below us another rocky terrace juts out toward the sea. People walk reverently along walkways paved in beach stones. We join them. At the center of attention is a smooth black rock set in a garden of yellow, purple, orange, and blue flowers. These are not the profuse and delicate spring wildflowers the poet loved, but Chilean nonetheless, and lovely. The stone in the garden marks the grave where Pablo and Mathilde are buried, their passage noted only with a modest plaque.

We discover that their remains were moved here from Santiago and reinterred on December 12, 1992, according to long-ignored instructions in the poet's will. In that will Neruda had also bequeathed his estate at Isla Negra to the poets of Chile as a place to reside and work, at no cost. The Communist Party was to be executor of his plans, which included new buildings and a dormitory. But Pinochet's coup stopped all that, outlawing the party and confiscating its property. Mathilde spent the rest of her life working to preserve the memory, the work, and the places her husband had given to the world. She died of cancer in 1985.

"What do we leave here but the lost cry / of the

seabird, in the sand of winter, in the gusts of wind. . . ." Neruda wrote in "Love for This Book," his poem to Mathilde in *The House in the Sand.* Even in translation, the last verse seems a fitting epitaph for him and for her, and for the story of Isla Negra:

It's late now. Perhaps
it was only a long day the color of honey and blue,
perhaps only a night, like the eyelid
of a grave look that encompassed
the measure of the sea that surrounded us,
and in this territory we found only a kiss,
only ungraspable love that will remain here
wandering among the sea foam and the roots.

TITICACA, ROCK OF THE MOUNTAIN CAT

I have been breathless and overwhelmed with beauty ever since we stepped onto the high ground of Cuzco in Peru a couple of weeks ago. South, behind us, were visits to Buenos Aires and Santiago, cities more European than any between Chicago and San Francisco, including all the towns in Montana. Bill and I had been traveling all night and were exhausted. We looked at the white city spread-eagled below. We looked at each other wide eyed. We did not speak. We knew we had entered the Third World.

Here is what we saw that morning: The highlands that had opened under our airplane's wings were patterned like a Japanese fan—terraced, green, and roadless; red earth had been planted in raised beds of corn and potatoes, and it rose again as the adobe walls and huts of villages. Snow-covered mountains encircled us, and deserts and jungles encircled the mountains. The air two miles high was so brittle it seemed glassine. We had landed among small brown-skinned people who looked at us with hard black eyes, and mixed with the familiar Spanish they spoke was an indigenous language resonant with ancient memories and secrets.

Bill and I had read in our guidebooks some of the stories told originally in Quechua. We are literate and sympathetic voyagers, but we knew right away that this was a culture we'd never be able to decipher at the heart. A mystery. Which, I believe, is as it should be, for no matter how much death the European conquerors brought, or what tons of gold and silver they stole, the interior reaches of Peru and Bolivia remain Indian—Native—impenetrable as Cuzco's great Inca walls, which stand solid as bedrock, never budging in recurring earthquakes that demolished the cathedrals and palaces built on their foundations.

Our trip to South America was a once-in-a-lifetime adventure, and Bill and I were bound to do the essential tourist things. Thus, when mudslides blocked the train to Machu Picchu, we dropped into the Urubamba valley on our way to the ruins in a Russian-made helicopter; we took a bus tour through the Incas' Sacred Valley, and were amazed at the dozens of varieties of corn; we even bargained for tapestries and hats and vegetable-dyed blankets in Quechua markets. But that was not enough. We wanted to go higher into the Altiplano, deeper into the culture. We had to see Lake Titicaca.

Our rattling, swaying, tenuously coupled train took

off from Cuzco on an all-day run to Puno, crossing a fourteen-thousand-foot pass through snow-decked peaks. The Altiplano (high plains) opened before us fogged and green, splashed with mud villages, alive with llama, alpaca, and the delicate white vicuña.

Dark-haired young women in red wool skirts peddled hand-knit socks and gloves at every stop. We ate big-kerneled corn on the cob with salty cheese bought from a peddler on the platform. Boys serenaded us with rousing Andean songs. As evening fell, we passed a trainside market that held a bewildering collection of nuts and bolts, tools and hubcaps, rubber and wire—indispensable parts of a do-it-yourself, poor, but industrialized society.

When we finally disembarked in the dark at Puno station, crowds of celebrants were drinking beer. Dressed in masks and sequined capes, feathers and fake fur, they'd been dancing in the streets. It was carnival time on the shores of Titicaca, but we felt too old and tired to dance that night. We gasped in the thin, damp air and took a taxi to our hotel.

The rainy-season February sky sits heavily on Lake Titicaca, and at 12,600 feet the clouds skim so low they seem touchable. They are strung like pearls over luminous daybreak waters—lemon gold and silver—precious metals gone skyward, it seems, along with souls of conquered Incas.

It is a haunted region, this netherworld of pre-Inca ruins and conquistadores and Franciscan bishops and Maoist Shining Path guerrillas, and us North American eco-tourists waiting for a bus. We are garbed in brown-and-white "baby" alpaca sweaters bought for a few dollars from street peddlers, and our wallets hang against our hearts, strung from the neck to keep them safe from poor and darker-skinned Andeans.

Exploitation has been the rule in the Altiplano of Peru and Bolivia since the twelfth century, when the Incas formed their empire. We like to think we're a gentler subspecies of exploiter, but how do you measure shame? A liberal like me should feel guilty.

"Remember the starving tin miners in Bolivia," was a cautionary tale my mother would use like Chekhov's little man with a hammer, striking me and my little sisters on the head every so often to remind us of our good fortune. She was trying to guilt-trip us into cleaning our plates of the lima beans, or the mashed potatoes in their congealed, buttery fat.

But I did not feel guilty then, and I do not feel guilty now. I feel an exaltation beyond history. Flutes sing like wind in the reeds. *El condor pasa.*

Titicaca is one of the highest navigable lakes in the world and the largest in South America, divided across its center by an invisible border separating the northwestern shores of Peru from Bolivia. Ancient cultures believed the lake was an inland sea connected to the ocean—mother of all waters. They believed this is where rain is born.

The name *Titicaca* means "rock of the mountain cat": *titi* being the Aymara Indian name for a species of wildcat who is believed to have swum from the mainland to Isla del Sol, where it was worshiped; and *caca* being the Quechua word for "rock." There is a Sacred Rock at Chincana on Isla del Sol, where the sun god Viracocha is said to have brought his son, Manco Kapac, and his daughter, Mama Ocllo, up from the deep to found the Inca line. An ancient legend claims that worshipers once saw the eyes of the Mountain Cat gleaming within the Sacred Rock. Thus the conjunction: Titicaca.

Places we encounter here, like significant places in every homeland, carry names that hold stories important to that place—history being little more than an accumulation of such stories. Koa, a tiny island off the northern tip of Isla del Sol, is named for the *ccoa,* a mythological flying female mountain cat whose eyes flash lightning, whose urine is rain, and whose thunderous roar is still feared by many Quechua believers.

One myth tells of a submerged city between the islands of Koa and Pallala. The story says there was no lake in those times, only dry land, and a temple forbidden to all except the holy virgins who served Viracocha. One day two men followed the priestesses to the spring by the Sacred Rock at Chincana, where they drew water. Surprised, one of the virgins dropped her vessel, breaking the jar. Angry Viracocha made the spilled water flow until it flooded the city and became Lake Titicaca.

Although divers, including Jacques Cousteau, have searched for the drowned city, it has not been discovered. But many artifacts have been found on a nearby underwater ridge, including stone boxes that may have been used for ritual offerings. Our Footprint series guidebook (a necessary companion in South America) suggests that a cult may have practiced human sacrifices in those waters to appease the feline weather god. Even today, fishermen setting nets for introduced lake trout or native pejerrey (kingfish) and karachi will not cross that passage for fear of angering *ccoa* and causing the deadly lake storms that kill, on average, four fishermen a year.

Somewhat tongue-in-cheek, the same guidebook advises travelers to be careful on Titicacan waters. In a country where violence and retribution are common in gods as well as humans, Aymara and Quechua traditions demand that anyone who falls into the lake will not be saved. They believe such unfortunates are fated to be offerings to Pachamama, the hungry mother of earth.

Bill and I had been enjoying the Festival of the Virgen de la Candelaria in Puno, but Lake Titicaca called from every window of our shorefront hotel. Our travel agent had booked us on a bus and boat tour to La Paz, Bolivia. We'd wanted to cross the lake on a hydrofoil, but he had chosen the Tansturin company's catamarans. We were not disappointed, for leisure over speed is usually the best idea. Who wants to whiz across the legendary waterscape at the top of the world when they can sprawl on deck for a couple of extra hours, pondering a geography-book fantasy imagined since the sixth grade?

The first leg of the thirteen-hour journey was by bus, a modest vehicle serviceable enough for our off-season entourage of Canucks, Aussies, and Trinidad Islanders. worried about taking off at daybreak on an arduous trip because Bill was recovering from a nasty fever-and-diarrhea bug, but the superantibiotic our Missoula health department nurse had given us was performing its miracle and Bill assured me he was fit to travel.

As we drove south toward Bolivia, skirting the reed-filled shore, past ever-present potato gardens and terraced plots of corn, our Aymara guide David explained the region's history. He pointed out several plain, adobe, evangelical churches, which, he said, attracted villagers because the more grand Catholic institutions had refused to teach the *campesinos* Spanish. Who knows, he said, what might happen if they teach the people the language of power?

Eighty kilometers down the road we made a rest stop in Juli, a colonial village with a washed and sparkling plaza and four Spanish churches. The red stone of the Church of San Juan Bautista had been carved and painted by seventeenth-century indigenous artists. Its flower and fruit motifs are typical of what is called the Cusquena school—a mixture of animist Inca art invading the austere European style. The fresh and appealing result is as strange as the melding of spiritual bedfellows it represents. About an hour later our bus crossed into Bolivia with minimal fuss, and we entered Copacabana, a handsome tile-roofed resort and fishing town on a peninsula that juts into Titicaca's southern quarter.

Copacabana is known for its early-seventeenth-century cathedral, whose sanctuary became a place of miracles after the presentation of a black wooden statue of the Virgin. Known as the Dark Virgin of the Lake or the Virgen de la Candelaria, she was carved in the 1570s by a grandson of the Inca Tupac Yupanqui and is the patron saint of Bolivia. When the Pope visited South America in 1988, Copacabana was one of his stops. It is a festival town—dancing and parades year-round—with a steep hill called Cerro Calvario, where pilgrims pray to the Virgin's priests for material wealth.

Leaving the cathedral, we passed streetside stands where the devout could buy toy-sized replicas of fine houses or boats or diplomas signifying college educations—emblems of desire made holy by association with the Virgin. Garlands of fresh flowers were also for sale, meant to be draped over actual cars and boats that had been parked at the cathedral's entrance. Their owners, perhaps nervous about their good fortune, stood watch by the gates, waiting for a priest to bless their vehicles.

I would have liked to stay a while, myself, eating fresh trout in restaurants, hiking the peninsula, sleeping in one of Copacabana's pleasant hotels, but it was past noon and time to embark. At the dock below town we boarded the catamaran *Santa Rita* and launched off toward the Isla del Sol.

On the foredeck the air was translucent, bubbly, imbibable as a pisco sour. The *Santa Rita* pulled out of Copacabana and I gave myself fully to oxygen-deprived ecstasy, a condition far preferable to the high-altitude headaches Bill suffers.

We were hungry after a long morning of sight-seeing and left the deck for a buffet lunch of regional dishes such as quinoa soup and the lake's flaky, buttery white kingfish. There were side platters of rice and vegetables, fresh fruits from the jungle, and to top off the meal steaming cups of *mate de coca.* This revitalizing tea, which we'd been introduced to on our first day in Cuzco, is a greenish infusion made from coca leaves that works better to mitigate altitude sickness than aspirin or other pain pills. Too bad it's not legal in the United States.

Soon after lunch we pulled into Yumani at the south end of Isla del Sol. Here, we transferred our baggage onto a ten-passenger day cruiser, which would transport us to the Bolivian mainland after our island visit. Bill felt a bit shaky, so he stayed at the dock with some altitude-sick passengers while the rest of us climbed a steep, intricately built stone stairway up the terraced sides of the island to an Inca fountain with three natural springs.

"Drinking of this one will bring you love," said our Aymara guide—a shy, large-eyed girl. The next spring promised health, and the third, perpetual youth. The West Indian men drank, but not me. At this point in my life the possibilities of loveless old age are preferable to the immediate possibility of dysentery.

Two paths diverged at the fountain. One led about a mile to the ruins at Pilcocaina. The other went up into the "ethno eco complex" owned by Transturin and built with assistance from the Bolivian state. Our guide led us up the company's path and we entered a beautifully designed and maintained hillside compound with a spectacular view of the scalloped island. Below us, boats ruffled the harbor's indigo waters. Cloud-decked glaciated peaks of the Cordillera Real crowned the baby-blue horizon.

Our way traversed terraced gardens named for Pachamama, which had been seeded with varieties of corn, potatoes flowers, and herbs used by Inca and pre-Inca peoples.A thatched adobe house, built in the local Aymara Indian islander style, was home to a family of five.There was a medicine and witchcraft center with dried and bottled herbal remedies, and outside, on a flat terrace, a woman in blue skirts, bowler hat, and magenta shawl worked a hand loom in the complex art of Andean weaving.

Below the Manco Kapac Lookout we wandered around a large corral holding fifteen varieties of llamas, vicuñas, and alpacas. Nearby, models of Lake Titicaca's reed fishing boats were displayed on an outcrop.The most famous of these boats were explorer Thor Heyerdahl's transoceanic vessels: *Kon Tiki, RA II, Tigris,* and *Matargui.* The island's shipbuilding Esteban family had helped create those boats, and one of the master craftsman's relatives was there to give us diminutive scale models.

The most impressive event of the excursion took place on the very top of the complex where, kneeling behind a stone altar—lake and islands, Mount Illampu and the Royal Range cloudblown in the background—a traditional healer called a Callawaya conducted a blessing ceremony for us travelers. Silhouetted against the sky, arms akimbo, wearing the typical Andean knit hat and red alpaca poncho, he was, despite my skepticism about such tourist-driven rituals, an arresting figure who evoked the power of a culture that is still alive and deeply rooted in Isla del Sol.

We had spent more than our alloted time on the island and our guide hurried us through the final attraction, which was an underground museum. Built in a cave, it was large (two hundred square meters), well ventilated, and not at all cavelike. Displays of the island's history were depicted in dioramas, mannequins, and maquettes. There was a fine collection of island artifacts and mummies, and representations of the Ekeko, or Abundance God, for whom the museum is named.

Ekeko is the symbol of Carnivale. His round figure may be seen in cities and towns all over the Andes. It is hung with miniature TVs and dishwashers, houses and off-road vehicles; dollar bills are pinned to his vest, as are college diplomas, candy, and hope. Streamers fly from Ekeko's hands, confetti dots his felt hat, and a cigarette hangs from his too-red, open lips.

Keeping Ekeko company, two larger-than-life carnival figures dominated the main corridor. The Devil, or Supay, was costumed in a red-horned serpent mask, sequined cape, and chartreuse satin trousers. Supay rules the underground and must be propitiated by all miners of the earth's treasures. Next to him was a black-faced Moreno—a paradoxical figure emblematic of African slaves and rich slave owners, as well as of the cruel foremen who worked the fabulous silver mines in Oruro and Potosi. More than two million men, women, and children—Incas, Aymaras, Quechuas, and Africans—were worked to death in those high, cold, satanic tunnels by the Spanish and their inheritors. The Moreno's clothes drip with coins, and he holds a cigar in his ballooning red lips. His eyes pop from strain and his tongue pants from altitude sickness.

Variations on these outfits, along with red-masked and wigged women embodying the devil's consort, white-garbed archangels, furry gorillas and bears, were worn by dancers we had seen in Puno. We would see them again on television, dancing in Bolivia's great Oruro Carnival on the Saturday before Ash Wednesday—one of the most lush and lively spectacles I have ever seen, making me want to go back to Bolivia if only to experience the real dance in real time.

The afternoon ended with a cruise from Isla del Sol through the narrow straits of Tiquina to the Transturin dock near Huatajata. Bill and I sat on the small deck, oblivious to any world we had previously occupied, dozing in the breeze, waking to sunstruck views of the white Cordillera of the Andes rising higher and closer as we approached Titicaca's eastern shore.

There was a run-in with local fishermen whose net our engines were shredding—waving of arms, shouting, and a speedy escape—and then sporadic conversation with an elderly Australian couple returning from Antarctica. I was once more amazed at the mobility of our kind, and appalled at the commodification of once difficult-to-reach exotic places (such as the lake we were traversing, or the Galapagos Islands or the frozen Antarctic with its ozone hole)—and yet we were as deeply into it as the retired Aussies, talking about coming back and staying longer.

This is the bourgeois quandary: how to take part in the rich possibilities of our times while trying to preserve what remains of earth's natural splendors and ancient cultures.

On the mainland we should have been warned by the sight of two grease-covered men in overalls fiddling with something under the tour bus waiting to transport us to La Paz. But the day was waning and we were tired. As the bus drove past villages and vacation villas along the lake, then inland and still higher toward the city, we spoke quietly or not at all, and paid no attention, at first, to the grinding of gears.

Passing across gravelly rivers fed by nearby mountains, I noticed tank trucks and pickups parked in the riverbeds, people siphoning and dipping water. Later, when our bus had decisively stripped its gears and we were unable to descend into the canyon where sparkling La Paz lay serpentine, the guide explained that those trucks were selling water to the poor who lived where we were stranded, on the high, barren, wind-swept Altiplano. This, he said, was Alto La Paz—half a million and growing—a new city of out-of-work miners and country people seeking a better life, many with no water or electricity.

Water, I noticed, gushed down into La Paz from the Andes in great open-ended pipes, bypassing the new city on its way to the old. Oddly, La Paz gets wealthier the lower you go—and warmer. It descends from slums at a chilly 13,452 feet on top to a mild bottomland canyon at 10,171 feet, where golf and tennis clubs and handsome houses cluster on a flowered, palmy, redrock desert.

While we waited for another bus to rescue us, wide-skirted, bowler-hatted women called *cholas,* their long braids tucked under shawls loaded with goods from the market, trudged up the steep, traffic-choked main road from downtown. A wire fence separated the toll highway where we were parked from the jammed streets, and as night fell, a man stopped on the opposite side of the fence from my window, unzipped his pants, and peed. I think he was making a political statement. Eventually we all crowded into a Toyota mini bus taxi and were dropped off at our hotels in the city's center.

We would rest in La Paz before flying home (a day of shopping, and a day of fuming because we'd been bumped off our Argentina Airlines flight), but our trip across Lake Titicaca glows in my memory as the actual end of our South American adventure—the incredible high topper. Months later, back in Montana suffering the pall of icy November, I can close my eyes and see the shining lake laced with reeds, the scudding clouds I might touch if I reached high enough. I see the Cordillerra Real standing guard over the lowland jungles. And viewed from the air through February mists, their glacial peaks appear as deadly and mysterious as the cat god's teeth.

MARGITSZIGET, WHERE IT ALL BEGAN

The language . . . oh, the language. I was in Budapest at last, my father's town, and everyone was speaking the language of my childhood. As I stepped onto ancestral ground, I was thrilled to see the great green plain of Hungary. And people walking easy in a liberty new and yet old as the city. But it was the language that gave me goose bumps. Hungarian is utterly unique, with no connection to other languages, except, perhaps, Finnish. Phrases lilt and swirl like Liszt rhapsodies; they crackle, dissonant as Bartok, in a cacophony of consonants. I tried to remember words, but what came back was a configuration of the lips. I loved the tongue-on-the-roof-of-your-mouth sound *tsz,* as in Margitsziget—the island in the Danube where I was heading.

I had come here for sentimental reasons. Both of my parents and all my relatives of their generation were

Hungarian: my father's family from Pest, my mother's from Translyvania (now part of Romania). Although my parents would become French nationals and, later, American citizens, and learned to speak both adopted languages, they never lost the accents of home.

The sounds of words, I believe, are as influential as meanings. Deep in my inner ear are the Magyar cadences of nursery rhymes and baby talk. When I went to prekindergarten, the children laughed, I'm told, because I spoke with a Hungarian accent. The way I write—the very shape of my thoughts—was molded by the broken English of my parents and grandparents.

In the Chicago apartment where I grew up my mother's mother lived with us, and my father's folks lived a few blocks away. The old people spoke Hungarian. My sisters Kathy and Carole and I were expected to speak only English (we were going to be *American* girls, not European Jews), and our parents spoke only English to us. But if we wanted to decipher the secrets the grown-ups told, or know what Grandma Beck was saying when she swore at us bad girls, then we had to listen closely, ask questions, and comprehend.

So from the moment of my arrival in Hungary, I was eager to speak the few words I remembered: *szervusz,* hello/good-bye; *igen,* yes; *nem,* no; *kerem,* please; and most of all *koszonom,* thank you—*koszonom szepen,* thank you prettily, than you very much for flying me home.

I had glimpsed Margaret Island—Margitsziget, in Hungarian—from the air, a lush patch amid the swirl of currents, a park resting like a verdant comma between the rush of twin cities. There are ten thousand trees on Margaret Island to catch the small May rain, and stone quays where men fish with long poles, and lawns as large as pastures, and yellow paths among shrubs of mock orange where old folks stroll under black umbrellas.

But we had not gotten there yet. The airport shuttle drove west through industrial suburbs marked by Soviet tenements, past a soccer stadium and a huge cemetery, then into Pest's old quarter. I was surprised by a sense of familiarity. Perhaps it was the midwestern feel of the air—moist and breezy; or the oaks lining three-story apartment blocks. Maybe it was streetcars clanging down the middle of avenues. Or men and women wide of cheek and wider across the beam. This was not the heady refinement of Paris, which I had left that morning. It was street life more common, like Chicago's.

I felt a surge of regret as I thought about my father, dead since January. I wished I had come with my parents decades ago, when they revisited their native country. Or with my Aunt Julia and my cousins. I wished my sisters were with me, my sons, all our children—for this visit should properly be a family affair.

My father grew up on the wide main avenue called Andrassy. His neighborhood included the ornate opera house, the Catholic basilica, and the Dohany Street synagogue. He had lived in a cramped courtyard apartment above the Jewish businessmen's club where his father was the caretaker and his mother served coffee. Now the only map that could guide me into the heart of his city was the emotional map of his stories—stories I'd had to pry out of him because, like many immigrants, he had come to America to reinvent his life.

I have always wondered why my family emigrated to Chicago. Why not New York? The answer lay at least partly in the land to which identity is attached. I believe Chicago attracted a Hungarian colony because, like Budapest, it is a city on a prairie, edged with water and sprinkled with great parks. Even the streetcars clanging down busy streets must have reminded the emigrants of home.

But not altogether. The monumental parliament buildings along the Danube are nothing like Chicago. Nor the quays lined with tour boats and casinos. And Chicago has no island. As I caught sight of Margitsziget lying green and low across the waltzing river, I knew I was heading toward a place and an experience entirely new, yet filled with keepsakes more precious than the usual snapshots, salamis, paprika, or embroidered trinkets that most tourists take home with them from Budapest.

Margaret Island rides plumb in the middle of the Danube, between the hills of Buda to the west and the sandy plain of Pest. It is a self-contained leafy world about a mile and a half long and five hundred yards at its widest stretch. Great pools and springs of medicinal hot waters lie beneath the island's northern tip and along the river's shores. I knew little about the island or its ancient history, and as the van passed over the Arpad Bridge, I was jittery with anticipation. I only knew I was going to the park where, after they were married in Paris, my father took my mother on an August afternoon in 1932 to meet his family.

In the liquid light, rowers punted past us along the island's western quay. Lovers kissed in the rain. The Grand Hotel rose in restored Hapsburgian splendor. I wondered if my parents had arrived by ferry: my father, Istvan (Stephen, called Pista), slim, dark, and soulful in an open-collared shirt. My mother, Ilona (Elaine, or Helene, called Ilush), diminutive, chic in her Parisian outfit.

The welcoming party included my father's parents, Johanna and Gyula Deutsch, his cousins Julia and Paul, his aunt Berta and uncle Adolph (the family's merchant "godfather," who paid the tab), and the Ermers from my grandmother's side. My father's joy at reunion was diminished because his brothers, Eugene and Albert, were missing. They had already emigrated to Chicago, where Albert, the adored first son, had committed suicide—a tragedy common among immigrants whose poverty and disillusion in the New World belied their great expectations, but never common in any individual case. Albert's death was still fresh in my father's mind and would become a shade hung over his marriage, his work, his children—a darkness that stayed with him until the end of his long life.

On the lawns of the Grand Hotel, in the shade of giant chestnut and plane trees, the family broke bread at tables draped in white linens. They must have feasted on goose liver, salamis and ham, stuffed green peppers, cabbage, Dobos torte, and poppy-seed cake. Perhaps the relatives toasted the new couple with champagne, but I like to imagine they lifted goblets of the ruby wine called Egri Bikaver (bull's blood), which my father would, years later, order by the case for our own clan reunions.

In the home movie of my mind I see my grandmother, short, stout, round eyed, smelling of lilacs; my grandfather erect, his face pocked and scarred by a skin disease he'd contacted fighting for the Austro-Hungarians in World War I, an affliction no doctor could ever name or cure. The dark-suited men would have embraced my mother in the courtly old-time manner—"*Kezet csokolom,* I kiss your hand." She, in turn, would laugh her bubbly laugh, charming them out of their socks.

All except my red-haired aunt Julia. "Julie came late," my mother had told me. "She'd gone to the ballet with a friend, and they sat at the corner of the table whispering to each other. Julie ignored me. Not one word. Didn't even look at me." Mother shook her head. "I never could forgive that."

Julia would emigrate to Chicago a few years after my parents. There, in a family scandal, she divorced her husband to marry her childhood sweetie and first cousin, my dashing uncle Gene, the ceramicist. They and their three daughters lived near us and, along with our grandparents, were the only relatives we saw frequently. We Deutchs (the *s* in Deutsch had been excised during World War II to separate our identities from Germany) were a tight little group made up mostly of women, except for the strong-minded and handsome artist fathers who dominated the pack.

A few relatives from both sides of my family had also come to Chicago, but we visited them rarely, for they were bourgeois—not artistic or political radicals like us. I'm sorry my parents (especially my increasingly antisocial father) participated in such intellectual snobbery. It would have been good for us girls to feel tied to a larger family, to know we shared a more common heritage. There were other relatives we saw occasionally, who had fled to Canada or New York, but those who stayed in Hungary were killed by the Nazis or simply disappeared. Which is why I had no names to search out in the homeland, no connections other than family memories and blood.

The Thermal Hotel, built in 1978, sits on Margaret Island like a layered Bauhaus cake. It has a lobby of tawny marble, and a display of modern art, and leather armchairs, and a well-stocked bar with a perhaps Communist barmaid who seemed to think serving well-heeled German tourists and North Americans was an insult not to be borne lightly.

The bellboy who opened the door to my room spoke passable English. I looked around, pleasantly surprised. This was no somber Soviet concoction but a large and airy chamber with a balcony overlooking a grove of cottonwoods, beyond them the gardens and quays of the island, and farther east the cityscape of Pest. Next to two smokestacks across the Danube, a neon sign flashed PALL MALL in red, white, and blue. I pondered a drink of gin, but headed instead to the hot pools—a more curative tonic.

Wrapped in a white terry-cloth robe large enough for an ample man, I descended to the ground floor. A maze of cubicles for treatments of ills ranging from rheumatic diseases to nervous disorders and chronic gastric catarrh led to pools, saunas, and a workout gym.

I disrobed in the ladies' dressing room, slipped into the swimming pool for a few brisk laps, then sank into the warmest of the three therapeutic baths.

The spa, like the hotel, was designed in cubist modules: Square baths looked out through square picture windows to a sunken terrace, potted palms, greenish filtered light. The mood was peaceful. Meditative. Nothing like the vivid art-deco decor, tiles, crowds, and chaos of the Gellert Baths in Buda, where the nude scene in the steambath caverns was worthy of Hieronymous Bosch and the outdoor wave pool was so powerful that an elderly gent nearly drowned and I was knocked to my knees.

The springs that feed the Thermal Hotel's spa flow forth at 158 degrees Fahrenheit and are cooled to between 89 and 104 degrees, too tepid for my proclivities but certainly soothing. This quality comes from sulfurous gases not strong enough to give off the odor of rotting eggs, but strong enough to be therapeutic. The gas in the water produces a film of bubbles that envelops your limbs like a silken second skin. I had hoped for effervescence, a champagne bath, but these bubbles were tiny and invisible. Many patients drink the mineral waters and take some home in bottles, but I settled for immersion only.

Each morning I soaked in the hot pools, sometimes after a massage and once following a mud bath. A pink-cheeked, cheerful woman slapped the mud over my shoulders, neck, arms, thighs, calves, and feet. She, like most of the hotel's staff, spoke German but no English, so our communication was limited to the universal language of the body. The thick, heavy mud had been transported from Heviz, mixed with thermal water, and kept hot in a kind of oven.

As I lay on my back like a piece of toast slathered in jam, tensions drained away. The attendant, perhaps in her midfifties, was singing a Hungarian folk melody. I thought about her life. She had been born at the beginning of World War II, when Hungary was allied with Austria and Germany. She would have learned to talk as Nazi troops took over the city, learned to fear as they were blasted out by the Russians. She would have cleared rubble, maybe marched with students in the 1956 Hungarian Revolution, and, after Soviet tanks crushed her city, helped once more in its restoration. Then came the "Hungarian Spring" of 1989–90. Goodbye Soviets, hello Eastern European refugees.

Today she tends German tourists, African American GIs resting up from Bosnia duty, and this Hungarian Jewish American woman smeared with mud. And all the while, I imagine, she's been singing the same song—a tune our grandmothers might have sung.

The springs below Margaret Island extend across the Danube and have nurtured life since the childhood of our species. Bone fragments believed to be nearly half a million years old indicate that Paleolithic peoples hunted reindeer, bear, and mammoth at hot pools along the river. Millennia later, nomadic Scythans and Celts settled in Buda; by A.D. 100 Roman Legions had built a fortress-city called Aquincum. During their three-hundred-year hegemony the Romans constructed a bridge to the island, and their ruins include fabulous baths with hot and cold pools. Then came the Huns under Attila, who drove out the Romans and united the three regions of Hungary.

By the end of the ninth century, Magyar tribes had ridden down from the Carpathians (a Christian prayer of the time was, "Save us O Lord from the arrows of the Hungarians") and created a feudal state. Eventually the Magyars converted to Roman Catholicism and their walled cities of Buda and Pest prospered, while the isle in between became a hunting preserve called Hare Island.

But peace did not last. In 1241 the Mongols invaded Hungary, killing about a third of its two million people. According to popular history, the besieged Hungarian king, Bela IV, vowed to commit his daughter to a Dominican nunnery on Hare Island if his forces drove out the Mongols. They did, and the nine-year-old princess spent her ecstatic girlhood as a bride of God. She died at twenty-nine and was anointed Saint Margaret of Margaret Island about seven hundred years later.

As Buda developed into a Renaissance center of trade and culture, Margaret Island remained a religious sanctuary with a Franciscan monastery, Dominican convent, and churches. But this too would end. The Ottoman Turks, who conquered Hungary in the sixteenth century, are reported to have converted the walled and turreted celibate retreat into a gigantic harem accessible by footbridges to encampments on both sides of the river.

In my vision of transformation, nuns like blackbirds with winged coifs give up their cells to belly dancers in diaphanous silken trousers. Sultans on Persian rugs take their pleasures within the cloister where a saint once flagellated her pubescent flesh. I imagine Oriental bells ringing from steeples in tune with the mating songs of birds, the cries of copulating rabbits.

Finally, armies of the Hapsburg Empire defeated the Turks, and the blue-eyed conquerors ruled from Vienna until 1867, when the Austro-Hungarian dual monarchy was established. Hapsburg gardeners planted thousands of plane trees on Margaret Island as bulwarks against the Danube's spring floods (one of which, in 1838, nearly wiped out Buda and Pest) and tamed the old hunting preserve with lawns and rose gardens, fountains, a thermal pool, and two hunting lodges stuccoed in Hapsburgian yellow. In 1873, when Buda, Pest, and Obuda merged to form the capital of Hungary, the first Grand Hotel rolled out its red carpet.

Victorian gentlemen and ladies, artists, poets, and musicians frequented Margaret Island, but the only way to get there was on a small white riverboat called the *Swan*. Soon half a dozen of these ferries steamed back and forth between the twin cities and the island's floating docks. The Romantic poet Janos Arany wrote his tender poems of old age, *Under the Oak Tree,* on Margitsziget. His marble bust joins the sculptures of cultural heroes such as Franz Lizst, Sándor Petöfi, Ferenc Molnár, Béla Bartók, and Zoltán Kodály along the tree-lined avenue called the Artist's Walk, which winds down the island from the Grand Hotel.

"Sainte Marguerite's Island," wrote a *London Times* correspondent in 1894, "pearl of the Danube, nest of flowers, sweet odours, and cool air, whence and whither the white steamers go—a spot unequalled by any one of the public gardens of any of the great cities from the Vistula to the Spree."

My mother's friend Zita is over eighty and lives in Ontario, but she grew up at the foot of Y-shaped Margaret Bridge, which has offered access to the island's southern tip since 1901. "Margitsziget was my playground," she had told me before my trip. "You had to pay a small entrance fee, which was enough to keep the proletariat away, and you could take a horse-drawn trolley if you didn't want to walk. On the way to the swimming pools or baths, you could stop to gamble at the casino, eat in open-air cafés, listen to gypsy music in the *beerstube* by the water tower. My father hated gypsies," she added.

There was an athletic club called Mutz, she explained, which was forbidden to Jews—even wealthy ones like her.

"We'd promenade on the *corso* along the Danube. Climb on the Roman ruins. Gundel's famous restaurant had a branch on the island in those days. My father took me there to dinner on Sundays, and once he hired a gigolo to dance with me." Zita laughed. "I was a girl. What did I know?"

In 1944, during the Nazi occupation, Margaret Bridge exploded at rush hour, sending cars up in flames and propelling hundreds of people to their deaths in the ice-flecked November Danube. The explosion was caused by booby traps hidden in the bridge's structure by the Germans, set off, perhaps, by accident. But it was no accident when, in January 1945, the Germans blew up the Grand Hotel on Margitsziget and the rest of the bridges across the Danube. They were retreating to Buda Castle, hoping to hold off the Soviet Army, which was advancing toward Pest from the east.

Most of my family had fled by then, or I would probably not be here to write this memoir. The vengeful, nearly defeated Fascists rounded up four hundred thousand Hungarian Jews in the last months of the war, sent many to Auschwitz, and summarily shot thousands of others.

"I was fourteen," an expatriate plumber from Arizona told me in the bar of the hotel. "We lived in Buda. I saw the Germans line the Jewish people up on the quay. Then they shot them. The bodies fell in the river. The blue Danube turned red. It was terrible. Why would anyone want to do something like that?"

Sunday morning. I have stuffed my tummy with a buffet breakfast (omelette and sausages, yogurt with sour cherries, nut-filled pastries) and am ready to embark on a last tour around the island. Sun shines. The paths are alive with strollers. It's a scene luminous as a Renoir painting, dotted with red balloons.

"Are there always so many children?" I ask an elderly bench sitter. She shakes her head and smiles. *Gyereknap.* Children's Day. That explains the games and crafts: balloon races, kids throwing pots on a potter's wheel, others on stilts or swinging from trees. There are stands with ten kinds of sweets, cotton candy, pretzels.

On the stage of a small amphitheater a blond puppeteer animates a dustpan with eyes, a pail with red lips. A young man with curly hair mouths words for a stuffed puppet villain in suit and bow tie. A cardboard bear wearing a pinwheel beanie holds out his arms like wings.

I see children and more children—in the rose arbor, under pines, playing in sandboxes, tended by parents and grandparents on the immense flower-rimmed lawn at the island's center. Girls in white stockings and patent-leather shoes. Boys in Chicago Bulls jackets. Teens with purple hair. There are toddlers at the musical fountain, small boys scouting for frogs among lily pads in the willow-swept Japanese garden. I head toward tents where people have gathered in long lines. They are waiting to give blood. Here, where almost everything is free, people are offering blood.

As I approach the Olympic swimming pool and the adjoining hot pools, water slide, and fountains, I am swarmed by a dozen or more whimsical pedal vehicles converted from bikes, some large enough for five, others like bumper cars with one or two madly driving kids. Families stand in line to rent them. I, however, am more attracted to a horse-drawn buggy decked out in flowers, the driver wearing a bowler hat, his white steed bridled in red-fringed, embroidered tack. If Mother were with me we'd ride in state like dowager queens, but I am happy on foot, and there is a smell of yeast in the air that I must follow.

The scent is coming from a *langos* stand near the swimming pool, a woman patting dough into a flat oval, then plunging it into a vat of bubbling grease. *Langos* (pronounced *lon-go-sshh*) is the fried bread we used to beg our grandmother to make on Sunday mornings. Salted. Dripping in butter. With or without honey. I buy a huge, golden round for a quarter.

Munching my *langos,* I walk past the thirteenth-century ruins where Saint Margaret was cloistered, then the tennis courts, a soccer field, and the balloon-bedecked bandstand by the Margaret Bridge, where dancers in yellow sequins are performing a perfectly silly samba routine—almost as silly as the nearby zoo, which features chickens, pigs, goats, and what seems the perfect Hungarian combination . . . a huge white goose harassing a jackass.

I turn back when I reach the monument that marks the entrance to the park. Unveiled in 1972 to commemorate the centennial of the union of Buda, Pest, and Obuda, the sculpture is a seashell (or, some say, the calyx of a flower) inscribed inside with symbols of Hungary's history. Designed by Istvan Kiss in the Marxist slant of the then-presiding government, the symbols are arcane enough to reflect a multitude of philosophies—Fascist, Communist, Turk, Catholic—as if to say, "Conqueror, take your pick." I can't help smiling. The artist's slyness is characteristic of a people who, at least in my family, are adaptive, creative, ironic, arrogant, yet hellbent to please.

It's my last night, and the Kiss Brothers (no doubt related to the Kiss sculptor) are playing gypsy violins, bass, and clarinet in the hotel's opulent dining room, which is filled with frolicking Germans, Hungarians, Americans, even a few French "of a certain age." They are dressed to the nines, sipping free champagne, feasting on chicken paprika, goose liver, cucumbers and sour cream, goulash, dumplings, and chocolate-filled crêpes called *palacsinta.* The food isn't up to my grandmother's or mother's standards, which, perversely, pleases me.

The Kiss Brothers are in serenade mode. They approach my table. "What can we play for you?"

I think of my mother, who should have been with me, friend and interpreter. Instead, she has suffered a heart attack and is grieving for my father in Chicago. I imagine her with skirts awhirl, dancing with my father and his relatives to fiddles, *cimbalom* (a gypsy dulcimer), and zithers in the summer of 1932, just across the way, on the lawns of the Grand Hotel.

"Play a *czardas,*" I say. "Play it for my mother."

I think back more than half a century to the living room in our Chicago apartment. Pinkish rug. Curved mahogany desk. Father's sculptures on the mantel. Gene's ceramic lamps glowing with muted light. Gypsy music on the phonograph. We would take off our shoes, join hands, and dance the joyful, stomping, circle dance. Mother, Father, little Carole hopping in pajamas, Kathy twirling like the ballerina she hoped to be, me awkward and wild, black pigtails flying—even Grandma Beck would dance, her long white hair flinging free from a tightly pinned bun.

The Kiss Brothers strike up the *czardas,* and though I feel silly as a donkey, I clap my hands as we did so long ago, and in my heart I am dancing the music of memory—here on Margitsziget—in the old country, where it all began.

I AM NOT THINKING

CHRONICLE OF A FATHER'S DYING

Here was the message: *I'll leave believing we keep all we lose and love.* One line from Richard Hugo's poem "Last Day There," copied onto a scrap of paper by my father and secreted in his wallet. Hugo, who died in 1982, only a few years after David, had become a friend of my father and one of his favorite writers. The wallet was my father's most personal possession. Like his watch and false teeth, he kept it at hand, always.

Now my father lay in a nursing home, naked as a newborn. Mother had taken the wallet home for safekeeping. It sat in his dresser among socks and shorts next to the Timex. I imagine my mother alone in their bedroom after sixty-five years of marriage. I feel her need to take that wallet into her hands, to stroke the cracked leather. I wonder how many times she picked it up then put it down, gingerly, as if touching fire.

Along with the credit cards, money, and organ donation cards, Mother found photos of women: herself as a young wife and in smiling middle age; me and Kathy and Carole as little kids and again as mothers; our two grandmothers; a moon-faced Korean girl; and Zee, the Georgia blond my father had married and divorced in midlife before remarrying my mother. Only women. Photographs *he* had taken. My father kept his harem literally in his pants pocket, a man who once said, "There is nothing I hate more than the sound of women laughing."

Each of us would tell a different story about the nature of his loving—all of them true. But this story is not about us, it's about my father's final passion. I believe he would approve such an investigation—arrogant and speculative as it must be—for on another torn sheet of white paper in his wallet he had written in blue ink a catechism to himself. It is a common spiritual idea, which calls out to me like a road map from the grave. My father had written: *The way is all.*

Easter 1996

Stephen Deutch, my father, sits in his chair—the new, padded recliner with the adjustable footrest, a piece of furniture common as a wart, unlike the Eames chairs, the Danish modern of our lives, each piece chosen like an object of art, upholstered by my mother with meticulous care. The recliner inhabits the room like a stranger.

He is a thin, wire-haired, large-nosed, gray old man,

hands clenched, face grim, clear headed much of the time. He cannot stand his inability to be self-sufficient, to walk away at will, control his bladder, or bring a fork to mouth. "No!" he shouts at my mother, as she tips a cup of water to his lips. "You're trying to choke me." He requires that she serve him just so. He can still order us about. We still obey.

My father has just come back from a long and painful rehabilitation in a nursing home. The injury was a broken hip, a fall, stumbling on his own foot in his own living room. Pain to him is old news. One knee has been replaced, and also his right shoulder—an operation done twice in his eighties, too late to repair the damage. My father walks haltingly with a walker, steps shuffling, balance gone. Pain blazes from back to thigh to calf when he moves, then eases, only to return after the next sit-down, the next effort to rise. The pain he suffers has many names: rheumatoid arthritis; pinched nerves; sciatica; a broken hip that mends slowly; old age.

For at least half of his eighty-eight years, my father's body has been damaged in one way or another. Slipped discs, broken ribs, aching shoulders, bad knees. He wore his scars like a tomcat on the fight—or so we believed, his aging daughters. We are likely wrong. He might always have hated his hurts, or better disdained them. No matter, his will to fight for movement and independence is almost gone. Or maybe the will is there, but the body will not obey. What counts is the letting go. He is giving in to pain at last. If he sits still, takes painkilling drugs, it may not find him.

As breezes drift off Lake Michigan whispering

promises of green, I know little of the long dying that awaits him, yet I know a transformation has begun, for my father's mind appears as broken as his body. While he was in the nursing home, Mother moved into a larger apartment, and the move compounds his confusion. The new apartment is on a lower floor. It has two bedrooms instead of one and faces Chicago's looming cityscape rather than the familiar blue harbor. Home must seem topsy-turvy as a dream to him. His art, books, furniture, sculptures—every object is here, but in a different spot, and where are his files, his slides, his keepsakes, the bits of string, the screwdriver he cannot turn with his swollen hands? He is frightened.

Fear is an emotion I do not associate with my father, because he always masked it with rage. In 1955 I brought a record of Dylan Thomas reading his poems back with me from college. On hearing Thomas's elegy to *his* father, *Do not go gentle into that good night,* mine adopted it as a credo. "When I die," he told my mother, I want those words to be read as my memorial." *Rage, rage against the dying of the light.*

Light, to my father, was more than a metaphor, it was the religion he taught us—his name for the vision that lay in a piece of raw wood to be sculpted, or an apparition in a doorway demanding to be photographed. All his life he had worked with his blunt hands to create objects of art. He would sit on a high stool at his workbench, tap-tapping with hammer and chisel. Heartwood peeled under his blade in curls, the scent of sap mingling with smoke from his cigarette. He had been a woodworker and art student in Budapest; a sculptor in Montmartre (where he supported his art by carving false antiques); then, tutored by my mother (a shooter for Paris *Vogue*), he learned photography, and they became partners in studios in Paris and, later, Chicago. Lighting was his specialty. "Look at the light," he told us.

If Steve Deutch had heeded his own advice, he would have been more fulfilled as an artist. He regretted time lost in commercial jobs when he could have been on the streets like Cartier-Bresson or capturing the faces of hunger like Walker Evans. Although his photographs were exhibited at the Art Institute and Cultural Center in Chicago, and nominated for a Pulitzer Prize by the *Tribune,* they were not exhibited in major New York galleries or published by a major New York press. The fault was largely his. He refused to hire an agent or sell photos through a national agency, resenting the fact that they got money for work he had done; and he was too prideful and shy to hit the streets as his own promoter.

After retiring from his photography business at the age of seventy-five, my father returned to wood carving, the art of his youth. A circle of admirers paid their respects, wrote articles, made TV documentaries about him, bought photos and sculptures. There would be two exhibitions and a Northwestern University Press book, but no recognition from the Great World. "I should have stuck with sculpting," he told me. "I shouldn't have wasted my time shooting catalogs for Marshall Fields for money."

Art versus Money. My father was tormented because he desired both. No matter how much he disdained it in theory, money was his most tangible mark of success and brought both comfort and joy. Rising far beyond the expectations of a Jewish janitor's immigrant son, he supported his parents, his mother-in-law, his wife and children, sometimes his brother, and often his eight grandchildren, with plenty left for his own security and their inheritances. The price of electing himself the family's patriarch was disappointed ambitions in art. He knew the price, and paid it—but not without a good dose of fury.

Now, as he sits crippled in his recliner, awake with eyes closed tight, light remains his password but blackness is his music. My father weeps often. He has always been prone to bursts of intense emotions, yet this is different. His face breaks apart. He can't bear to see his wife sign the checks, or to sit by while his children negotiate the power of his money with his lawyer or stockbroker. Raging at manhood lost, he weeps in frustration and shame and wishes he were dead.

In fact, he is very much alive. And kicking. Sometimes literally kicking at my mother. Anger, we can respond to, but it is hard to deal with blankness. When depressions come down on him, as they have so often these last years, my father lies on a couch, rigid, hands at his sides, not speaking, eyes shut.

Music is the path that can lead him out of self-imposed darkness, Bach or Beethoven or Verdi flooding the room with waves and patterns of sound—music his great solace. But lately he has been choosing the dark when he is not depressed, closing his eyes even while eating. Where is he going, I wonder? What is he thinking? In this state even music does not touch him.

He is consumed, I believe, with a passion that cares nothing for the light of the worldly world. He has serious work to do in the interior dark where none of us can intrude. He knows, has known for many months, that dying will be the final artistry.

Labor Day, 1996

My father sits in his new recliner in the apartment that is home but not home and looks at the walls. His back is to the window. I try to draw him toward the view, sun setting red in the west, earth and sky shining in hues of orange and apricot, skyscrapers aglow like canyons at sunset—the slanting play of light he has always loved, but he cannot be bothered to turn his head. The television is another window he ignores, preferring to look straight ahead at a watercolor painted by his late friend Aaron Bohrod. Chicago in the 1930s, the Chicago of his thirties: a butcher shop, a milk wagon, the curving street, a windblown grocery boy, the pale blue winter sky. Reflected in the painting's glass like a double exposure are images of smoke rising from a chimney across the street, fiery clouds, shadows in motion of ourselves.

Here. This old woman. This white-haired daughter. These walls covered with art and artifacts. His sculptures of wood, his dead brother's swan-shaped bowls. Shadows. Reflections. Here is what his world has come to. He closes his eyes.

There was once a planet full of surprises to explore.

A voyager who spoke three languages, my father crossed the Atlantic when he was thirty to settle in Chicago, but that was just the beginning. He would travel to Mexico and the Deep South, to the Rocky Mountains and South America, into West Africa, Poland, Turkey, Japan, taking pictures in every place.

On the track of an image, he was intense—engaged in seeing. He looked like John Garfield in a 1930s movie. Rolleiflex slung on his shoulder, my father roamed back roads and slums, exposing hidden lives, capturing the ordinary as extraordinary. I was embarrassed at the boldness with which he approached black worshipers in front of a storefront church on Chicago's South Side or housewives hanging laundry in a Naples alley or peasants selling apples in a French village market, how he enticed them into familiarity. It seemed a way of soul stealing, which made me cringe. But the farmers threshing wheat in India did not cringe. Nor the Greek crone with her donkey. Or truck drivers in the Sahara. They invited him and my mother into their homes, or to celebrate their weddings, understanding that the attention he paid was a kind of homage.

My father's walls are hung with trophies of his voyages; a carved mask from Senegal, a Penitente crucified Jesus from Chimayo, Japanese figurines, a disjointed Italian puppet. He looks at his walls (his world these invalid days) and the art comes alive, acting out stories as real to him as we are. Dead faces return in vivid motion—my red-haired Aunt Julia, his brothers Gene and suicidal Alfred (dead for sixty-six years). An embroidered white horse he brought home from Arabia thunders toward him from its niche on the knickknack shelf. The cubist shape of a heavy-breasted woman emerges from a painting and conspires with a bearded man.

Strangers invade the house of his mind. They threaten, they ignore, they are in the next room, stick figures with hats. He will not let go of their stories. He calls 911. He wheels down to the building's office, accuses the manager of opening a secret passage under his new recliner. It is through this hole that the bizarre intruders fly into his living space.

We listen to his tales in a patronizing way. We indulge him as you would a crazy person. He knows this, but we are his only audience. "I've figured it out. They're a troupe of actors," he tells me with relief. "They're rehearsing in the living room."

Good. This version of his obsessive hallucination is not threatening. The stick figures with hats are actors, not kidnappers come to steal my mother or murder us in our beds.

"It's just your imagination," I say. "Like a dream. Those people don't really exist outside of your head."

My father looks away. The pleasure of confiding in someone who might understand is gone. His face becomes blank and hard. Arguments will be of no use. Like all of us, he believes what he sees. Arguing simply creates distance between us. I try a different tack. It is four in the morning. He has gotten out of bed and is stumping around on his walker. We sit on the couch in the living room.

"Look," I say. "I think I understand. You are an artist, aren't you?"

He nods his head. "Yes. An artist."

"You're an old artist who can't practice his art anymore."

Again, yes.

"But you want to?"

"Of course I want to."

"So maybe you're making art in your head. It's more interesting than your real life. Maybe you are bored."

"Maybe." He smiles. His smile these last months is sly, like the grin of a small boy caught doing something naughty, a spontaneous delight in trickery. "Gotcha!" says my father's grin, and I realize he is indeed part trickster. Yes, he is bored. Bored since he retired from business, from travel, from every meaningful activity except wood carving and family. He's been indifferent to his life, I guess, for ten years, maybe twenty.

Okay. Now I comprehend what he was up to in some of his later sculptures—the hastily crafted Kama Sutra figures in positions of cock sucking, penetration from the rear—and his round-bottomed women dancing around, actually worshiping an erect, black, polished penis. These are more than anthropology, or the fantasies of a randy, impotent old man. They are jokes on his wife and daughters and pious admirers. They are, as he said of his final photo and sculpture exhibit, "my last hurrah," a tip of the hat to the primal forces that inspired him.

Look, his smile tells me, I have imagined and rendered suffering in my Holocaust figures, my *Pietà,* my starving Rwandans. I have tried to translate the musical fluidity of wood into carvings of accordion players, violins, dancers. I have made a whale and a death mask and a woman lusting for a man's penis. I can try to create anything I damn well imagine. Anything that moves or delights me. These visions are no different. You can accept my stories or go to hell.

After sixty years, I am able to smile with him. No backing away because of prudishness. I realize I have denied myself intimacy with the most salty aspect of my father's character out of my fears of his sexual power. I was looking for a high-minded model to follow, a hero not a coyote.

In this I was influenced by my mother. Mother always insisted her husband had no sense of humor. Laughing is her hallmark, not his—a lightness of spirit that draws him even as he abuses her. She is the bright water to my father's mountain, a Yang for his Yin, but she weighs him too heavily.

"She's so beautiful," he confides to me as Mother waltzes into the living room in her powder-blue sweat suit—a ninety-year-old, humped, white-haired women, cheeks pink, eyes bright from a walk in the frosty streets. "Sometimes when I look at her my heart is so full I feel like it will burst."

Thanksgiving, 1996

"Annick is here," my mother announces as I arrive once more from Montana.

My father seems to be sleeping. He rouses himself, turns his mouth up in a faint welcome. "Hi, Annick," he whispers.

I'm still not used to this whispering voice, which, like his life, goes dimmer and dimmer. I wonder if the loss is due to muscular disability, neurological failure, or a willed refusal to speak, but I do not wonder long, for tonight there is a fourth person in the apartment, a Polish man in his twenties.

Mother hired him in September, shortly after I left. A day arrived when she could not lift my father from his chair. She could not bathe him, change his diapers, apply the catheter to his penis at night, hoist him into a wheelchair, drive the car. The Polish boy has been a godsend, like a grandson to my mother. Father refuses to learn his name.

"Mariusz," I say. "Mar-ee-ush."

My father turns away. He thinks there are two boys. He doesn't like either of them. He calls Mariusz "that guy." But he is glad to see me, more animated than he has been with others, or so the psychiatric visiting nurse will tell me.

Psychiatry is much on our minds. One of my jobs this week is to drive my parents to doctors' appointments. The half-hour trip to their HMO medical center in Skokie has become the only way of getting out of the house for my father, since the weather has turned cold and he does not want to go outside in his wheelchair. The psychiatrist is a large, ruddy, bearded man.

"I like you," says my father, and they laugh. "I wish I could die," he says. "I will send you my book for Christmas."

My father has been taking Haldol, an antipsychotic drug. Under its influence most of his hallucinations have disappeared. But the drug increases his immobility, makes him drowsy, swollen, stiff as a zombie. The psychiatrist prescribes a new drug that will not have the zombie effect. He takes pains to explain the symptoms of geriatric dementia to all of us, including my father.

"It will only get worse," he explains. "Steve is like a three-year-old child. He wants what he wants, now. He will throw a tantrum. You cannot reason with him, like you can't explain to a child why it's not a good idea to eat cookies before dinner. Just say no. Separate yourself from him. Walk away."

The doctor's explanation is simple and useful. We believe it too well, yet the old habits are hard to break. So one dark afternoon when my father is struck with one of his obsessive delusions—he has a job he must do, a job of photography—I try to talk him out of it. But my father is adamant. "I've got to go to the construction site," he says. "I must see if the tank is there. A used tank. And a pile of gravel, like a hill."

He has mentioned this construction-site job several times and we were able to deflect him, but not now. "I'll go myself," he says, pushing himself out of the recliner and onto his feet.

Mariusz hurries to help him. "Where are my boots?" my father demands. "I want my boots. You've stolen my boots!"

"No, here they are." I pull his work boots out of the closet. "We wouldn't steal your boots."

Mariusz on hands and knees tries to jam my father's toes into the boots. His feet are too swollen to even fit halfway. "All right," says my father, "get my other shoes. I can see what I need to see from the car."

Mariusz looks at me, eyebrows raised. I shrug. Why not? It'll be good to get out of the apartment. And so we bundle him up, scarf, gloves, wool cap, and jacket, push the wheelchair to the curb, lift my father's deadweight body into the front seat, limb by painful limb.

I take the wheel. "Where to?"

"Addison," my father directs. "Go west on Addison."

"What's on Addison? You want to see Wrigley Field?" I had thought of driving downtown, or to the Planetarium (my father's favorite view of the city).

"Just go where I tell you," he orders. So we drive west on Addison, past an endless run of shabby brick apartment houses, small shops, bus stops, fast-food joints, and mini malls, west into the city's vast interior. A thin snowfall dusts the grayness. I want to turn back, but my father urges us onward. As we approach the Expressway, he commands me to turn left into the parking lot of a Kmart shopping center. "Keep going," he says, and though I am sure he is mad, I drive along the edge of the lot toward a deserted railroad embankment. "Stop here."

There, spread like a miracle behind a wire fence, is the construction site he has been looking for. A pile of gravel. An old Caterpillar excavator that could stand in for a tank. Tall, rank weeds crowd up the embankment, which is littered and flagged with plastic debris. An afterglow of the sun throws the scene into backlit silhouette against a dim yellow aura of light.

My father studies the place intently. "Drive over there," he commands. "I want to see it from that angle."

I do what he says without question or judgment, for I am filled with awe. In the backseat Mariusz mirrors my excitement. How did my father come to know this place? He hasn't driven his car for months. And besides, the site is hidden from the street. You wouldn't notice it in passing. Something mysterious is happening. This flying directly toward a previously unseen destination of the imagination seems akin to the radar built into birds. After a few moments, my father nods his head. "It's not perfect," he says, "but I can work with it."

"We've got some time before dark," I pipe up, eager for more miracles. "Is there somewhere else you want to see?"

He points toward the Loop. "I'm looking for two hills."

I drive down working-class streets in search of higher ground. We pass a Greek Orthodox church, its spires like two peeling, gilded onions. "There," says my father. "Those are the shapes I want. I need two hills like that, and there will be soldiers on the top. Two soldiers, maybe three, on each hill. I will shoot them with their backs to the light. In wind. You will not see their faces, but they must have uniforms in rags. That will be the second photograph. The two go together. The used tank. Hills with soldiers in ragged uniforms. It's about war," he says. Then he corrects himself. "Postwar."

I can see the pictures in his mind, black and white. Complete. Powerful. Postwar. My father has not been hired to do this job, but that is of little importance. What matters is he has a job to be finished, a job he could not complete without going out to see what had to be seen—real objects in the particular neighborhood.

As we head home in falling darkness, my father is quiet, satisfied, at rest, and I am ashamed. The world, as Walt Whitman said in an unfinished poem—"the earth and the hard coal and rocks and the solid bed of the sea"—is positive and dense and actual. It is not a joke, nor any part of it a sham.

Terrible Waters

I do not usually keep a journal, but one night I drink too much of my father's bargain-basement Scotch whiskey. I scribble a few lines.

> *11/26. A bit drunk. Anger confirmed. You don't really like your parents. Father the authoritarian. Child in him is suspicious, mean—manipulative. Mother—doggedly stubborn. The overfeeder. The person who must be fed. Me, rebellious. Can't stand being controlled by strings of guilt, duty, obedience. Brings out the cold heart, the predator eye.*

On the final night of my visit, my cousin Nancy comes for dinner. During the afternoon my father, mother, and I had gone to the Merchandise Mart with a young artist, one of my father's disciples, to see an exhibition of his paintings. By dark we are all tired, but I am happy to see Nancy. She is a jolly woman, stout and round eyed as my grandmother Deutch, my father's mother, called Honey (short for Johanna). And Nancy reminds me of her father, my uncle Gene—the elder brother who worked with clay, joked, drank beer, and tossed me in the air as a child—a man who loved life and died of liver cancer at fifty-five.

My father has always been drawn to Nancy, and he holds her hand at dinner, flirting in the old way. Fed by Mariusz, he wolfs down my mother's chicken paprika and homemade dumplings—eating the only bodily pleasure left to him. A man addicted to sweets, my father has two helpings of pound cake and strawberries topped by Cool Whip. Not once does he close his eyes.

When he is finished, I pull him from his chair and place his hands on the walker. We all stand up from the table, and as I attempt to help him towards his recliner, he roughly shakes my hands away. "Not you!" he barks. "I want Nancy!" I back off, hurt. He smiles his nasty, sly smile.

When my cousin departs, we watch television as usual, but my father does not nod off, as usual. There is an old film noir playing (like me, he loves murder mysteries—the darker, the better) and he attends every moment of it. Then, after eleven and past his bedtime, he refuses to go to bed. We are all exhausted and somewhat exasperated. I have to get up at six in the morning to catch a plane to Tulsa. I sleep on the living room couch. I want a few hours of rest.

When my father finally falls into sleep, we settle down for the night. About 2 A.M. he wakes up. His breathing is labored, phlegm rattles in his throat, but that is not unusual. He has smoker's emphysema. I give him water, a pill to relax him, plump up his pillows. Mariusz in his briefs stands blinking in the doorway. Mother, who sleeps in the same room as her husband, on a twin bed a couple of yards away, pulls her covers over her head. My father lies down again. We lie down again.

"Help me! Help me!" A call in the dark, cried over and over in monotone, wakes me within the hour. Shocking, this piteous plea from a man who all his life refused to ask anyone for help. Now all of us are up and alarmed. My father has a pain in his side. I take his temperature. No fever. He wants a hot compress on the hurt. Mother makes him one. He wants warm milk and honey. We bring it. The pain, he says, is bad. Eight or nine on a scale of ten. He will not rest. When he demands chocolate ice cream, that's the last straw. His psychiatrist had said to treat him like a three-year-old.

"Look," I say, stern as a bad nurse, "there's nothing we can do about your pain. You don't seem sick. You don't have a fever. You're being inconsiderate, keeping all of us up. Just lie down and be quiet."

His eyes scan mine with a kind of loathing. "I would never believe you could act like this. For the first time in my life," he says, "I want to slap you in the face."

I sit beside him on the bed. Put my arm around him. Tell him I'm sorry. That I know it must be terrible for him to be so helpless. That we, also, are helpless to ease his pain. Not any specific pain, but the hurt of being old and sick and dying. He relaxes a bit. Sighs. Lies back on his pillows, closes his eyes, and falls instantly into a deep sleep.

Now it is four in the morning. As I toss on the couch waiting for the alarm to ring, I hear my father's labored breathing accompanied by low moans. I do not know if the moaning is real or another play for attention. I am troubled and angry. Hurt by him and hurting him. I cannot wait to get out of there.

Next evening I call my parents from Tulsa. The machine answers in his halting, accented voice: "This is Stephen Deutch. Sorry we are not able to answer your call. Please leave a message and we will call you back as soon as we can."

I am worried. I call Carole in San Francisco. "He's in the hospital," she says. "Pneumonia again. They couldn't get him to wake up. Then he collapsed in the bathroom. The visiting nurse heard pneumonia in his right lung. They called an ambulance."

Have you ever felt your heart squeeze with shame? Blood flushes through your body, cold, like a river at flood. "Oh God," I say. "He was right. I deserved to be slapped."

"Don't be so hard on yourself," says Carole. "We've all been making mistakes with him. He's not who he used to be. You don't know how to act. Don't know what's right to do."

"It's strange," I tell her. "He's known all along. Last September he said, 'I'll die from pneumonia.' He wasn't alarmed or anything. Just an offhand remark."

I wonder why it is so hard to be compassionate. To treat a person consistently with kindness. To believe their hurt is real. Compassion was not easy for me when it counted most. It was the hardest thing in the world that night with my father to put aside selfishness, resentment, anger, and hurt. To forget myself while caring for one of the few I love deeply.

Here, then, is my confession. There is no way to erase my guilt, no redemption in future acts of good faith. The coldness of heart that caused me to hurt you, Father, is a reminder of my vulnerability and our mutual humanity. Like Odysseus crossing the River Styx, I had entered the world of death and was not prepared for the horror. "It's hard for the living to catch a glimpse of this . . .," said the shade of Odysseus's mother. "Great rivers flow between us, terrible waters."

Christmas, 1996

I arrive in Chicago a week before Christmas to oversee my father's move from hospital to nursing home. It's crisis time. In less than a month he has had two bouts of pneumonia; Mariusz has gone back to school in Poland; Mother is alone and distressed with our good Doctor Stanley, who has thrown up his hands.

Father's pneumonia is under control, says the doctor, but will never be cured. He is plagued with brain seizures that cause him to blank out for hours, and the medicine to control the seizures acts like a sedative. Besides, he is not allowed to eat solid foods—the last pleasure gone—because he inhales chunks into his lungs, causing more pneumonia. An IV drips antibiotics into his veins. Because his condition is incurable, the HMO will not pay for more hospital care. Medicare will fund a few weeks of rehabilitation at the nursing home where he stayed while his hip mended, but only if he makes progress. That nursing home is my father's hell, dying there his worst nightmare. I wonder what kind of rehab is possible for the dying.

The man I see in his hospital bed is thinner, weaker, more helpless than the father I said farewell to a month before. He acknowledges me, as always, with a smile and a whispered, "Hi, Annick." The smile says everything. He has forgiven me, I hope, not merely forgotten my trespass. He tongues a spoonful of mush, grimacing, then lies back, toothless mouth open, in a trance. I am frightened by his stillness. My father does not move as the ambulance guys lift him onto a gurney, wrap him in white blankets like a mummy. He does not twitch or flick an eyelash.

At the nursing home he still does not move, but lies rigid, lips agape, eyes shut tight. The nurses want to send him right back to an emergency room, another ambulance, another hospital, but I will not let them.

"He's not a Ping-Pong ball," I say. "Let the poor guy rest." Mother and I are appalled. How could our beloved doctor warehouse a man so obviously ill?

I make frantic calls. Doctor Stanley interrupts his Christmas vacation to call back. This is the beginning of the end, he says. We must prepare ourselves. Years before, my father signed a living will saying he did not wish to be kept alive by artificial means. He wanted no tubes, no machines. We should let him go his chosen way. The IV will be detached in two days. If my father starves himself, death might come in a couple of weeks. If he eats, the dying could last a month or two. In any case, the doctor assures us, he will not be suffering pain. The body defends itself as it dies in this natural way of dehydration. It becomes numb as a seashell.

The staff at the nursing home wants to force-feed my father through a tube down his nose. They are trained to preserve life at all costs. Mother, as the executor of his living will, must sign the order to prevent forced feeding. She must sign her husband's death warrant, and I have to help her do it. Such a decision is an agony beyond description. We will do nothing so drastic without my father's consent. He is not comatose all the time. Often, his eyes are open. Sometimes he speaks.

"You're not eating enough to sustain yourself," I say to him. "They want to feed you through a tube down your throat. Do you want that?"

My father looks up at me. He shakes his head. He mouths the word *no.*

I will never forget the tenderness of my mother during the final days she spends with my father. How she sits for hours, moving only to wipe his forehead with a moist towel or softly comb his wiry bush of gray hair. How, when I coax her away at night, she will bend over him, having to stand on tiptoes because she is as small as a child. She holds his head in her hands, caresses his cheeks, kisses him so sweetly on the mouth. Such moments are too intimate to watch. I move behind the curtain that separates his bed from his roommate's bed. I wait until I hear her whisper, "Good night, Petit"—*petit,* little one, their pet name for each other, dating back to courtship in 1920s Paris.

At home my mother weeps so hard she cannot talk on the phone to my sisters or her best friend. She cannot sleep, even with Tylenol PM, haunted by the visage she calls a death's head. She tells me she will not be able to bear watching her husband die. "He broke my heart once, and now he's breaking it again," she says. "But when I look at him like this, I'm glad he had some fun. I forgive him. Maybe I should have had lovers too."

On Christmas Eve I cross paths with my sister Kathy in O'Hare Airport. She and Carole and I have been rotating visits and care for our parents for over a year. We are changing the guard now, me going home for a few days with my family in Montana; she coming from Boston to attend our father, and to help our mother through this sad Christmas. As it turns out, Mother will not share her husband's last journey. Two days after Christmas, while I am snowbound in Montana, she suffers a heart attack brought on by stress and grief. Luckily, Kathy and her daughters Alison and Jessica are on the spot. Mother weathers the heart attack, but is barely safe in the cardiac care unit when she is attacked by pneumonia and spasms of the larynx that choke off her breath.

The call comes as I am shoveling snow with my sons—four feet on the ground and more falling—the cedar roof, the log walls buried in white, our road plowed out with a half-track cat. Kathy is calling in emergency mode a second time. Two nights ago the heart attack, now this. "She's choking," Kathy weeps. "They can intubate her [put her on a breathing machine], but the cardiologist says she's so old we should let her go."

"She's not ready to die," I cry. "We're not ready to lose her—not both of them."

The young Chinese heart doctor, a family friend, comes on the line. We must make a decision. Now! "What would you do if it was *your* mother?" I ask.

"I'd let her die quickly," he says, his voice soft and concerned, but sure.

I stand silent, the tears streaming. "I can't do it," I say at last. "She's not like my father."

While we hesitate, Mother has demanded the tube down her throat. She has decided to live. Kathy comes back on the line. "She's breathing," announces Kathy. Saved. Perhaps for more grief, but grief is also life. We order pink carnations, red roses, purple iris, ferns—a get-well bouquet fragrant with hope.

New Year's, 1997

On New Year's Eve I join Kathy and Carole in the Chicago apartment, which is bereft of parents. The year turns and we children are in charge for the first time in our lives. Mornings, afternoons, and evenings we drive the icy streets from Mother's hospital to Father's nursing home. Each day may be the last day. Each moment the last moment. We are saying good-bye to our father, get well to our mother, and squabbling, jockeying, laughing as sisters will, holding tight to one another.

While Mother learns to breathe again and swallow food, my father lies immobile, starving himself. His brain seizures and chronic pneumonia remain uncured, but when we enter his room on New Year's afternoon, his face seems smoother than it was a week ago, calm and pink. His breathing is soft and regular. There is no agitation to his skeletal body, which is moved twice daily from one position to another by the strong Asian and African American women who care for him with a devotion near to love.

My sisters and I have brought Mozart sonatas and Beethoven quartets, Puccini arias, Sephardic folk songs, and the pure abstract beauty of Bach's cello suites performed by Yo Yo Ma—the music my father has asked my mother to play at his funeral. Music has been his joy and solace since childhood. When he was nine, according to family myth, he dislocated his shoulder while ghost-conducting Verdi in an alley behind the Budapest Opera House. I remember a Sunday morning, the hi-fi at ear-splitting volume, him singing and weeping with Rigoletto, the tragic clown.

Our father seems to be sleeping. We put Bach on the boombox and turn up the volume. Melodies radiate through him, through us, and out into the yellow halls among old folks in wheelchairs, white-coated nurses, aides in their rainbow smocks. We whisper to each other. We hover around his bed. He opens his eyes and the peace of his visage is shattered. He knows it is holiday time. We should be in Boston, San Francisco, Montana with our families. But there we are, three gray-tressed, red-eyed girls rousing their father from sleep, kissing his forehead, touching his swollen hands.

My father looks from one daughter to the other, more bewildered than pleased. He opens his mouth. He speaks the first of the last complete sentences we will hear him speak.

"Why are you here?"

The answer is obvious. Not good news. He turns away with a frown. Shuts his eyes.

Two days later there will be a second sentence. He has been getting weaker, but still struggles to choke down a few spoonfuls of applesauce or Jell-O. He swallows his bitter medicines. I watch his eyes move from my face toward the bedstand, then back and forth again, searching. I bend to hear his hoarse whispers.

"Where is my money?"

"Where is Mommy?" I echo, not trusting my ears. We had told him Mother was ill. But we had not told him that she too was near death. I imagine he has guessed the worst. He moves his head a breath. His eyes are impatient. Laboriously, he mouths one word again.

"Money."

Money. I nearly laugh. These are not deathbed words of wisdom. I tell him his wallet is safe. It's home in his drawer, along with his watch. He clamps his mouth in a grimace of displeasure. Shit, he must be thinking. They have stolen my wallet. His money is gone, the emblem of power. It's over.

I have been rereading the *Odyssey* this winter, thinking about journeys and death. The classics scholar Bernard Knox, in his introduction to the Fagles translation, explains a passage from Book 11, where Odysseus crosses the River Styx and enters Hades. He wants to converse with the dead, but ghosts do not talk. To restore to the dead the gift of speech, Odysseus must make elaborate animal sacrifices. Blood must be shed.

"The shades crowd round the sacrificed animals," says Knox, "yearning for a draft of the blood that will for a moment bring them back to life, restore memory and the power of speech."

Perhaps after we are dead, our souls will want to speak. Life after death is conjecture, but the process of dying is real and well documented. Memory usually flees first, then speech. Natural death is marked by a turning away from the living, a turning inward toward reconciliation with the self. This is what the hospice nurse tells us when we finally call for succor. This is what our father tells us before he sinks forever into silence.

We enter his room on the next-to-last day and he is wide awake. His brown eyes are dense and hard as buttons. Restless eyes, they move from our faces to the window, out to where snow is falling. We stop our chatter.

"Do you want to say something?" asks Kathy.

"What are you thinking?" asks Carole.

The question hangs heavy in the silent winter room. At last my father opens his mouth. He does not whisper, but speaks clear as any ordinary mortal.

"I am not thinking."

We look at each other, amazed. Our father is fully aware of his condition, his journey. He will not speak again. "I am not thinking" is the cryptic farewell he offers to his children.

Last Day

I wish I could, in full faith, cry out to some god for help in this grief, but I can only cry out to the voiceless sky. As I watch my father die, the death of my young husband thirty years ago returns to sadden me. I ponder my own death, wondering why I cannot put words to my inchoate yearnings. Luckily, the Polish poet Anna Kamienska has made the prayer that I would make.

A Prayer That Will Be Answered

Lord let me suffer much
and then die

Let me walk through silence
and leave nothing behind not even fear

Make the world continue
let the ocean kiss the sand just as before

Let the grass stay green
so that the frogs can hide in it

so that someone can bury his face in it
and sob out his love

Make the day rise brightly
as if there were no more pain ...

On January 6, I look in the morning mirror and see that I am pale with grief. My eyes are yellow as piss holes in the snow. "Go home, Annick," says Carole. "Take a vacation."

I have been too long among the old, the sick, and the dying. It is time to fly to my Montana mountains. We have just learned that Bill has prostate cancer, and I must attend to loved ones at home, who need me as I need them. I know I must go. But first I must say my goodbyes.

Once more my sisters and I cluster around my father's bed. He lies in the fetal position—curled and at peace, brow smooth, the furrows of pain deleted. We catch our breaths, astonished. We have never seen him in this position. For as long as we can remember, he slept on his back, chin tilted upward, arms at his sides. The arthritis in his shoulders was too painful to bear weight, his spine too rigid for bending. On this, his last day, our father sleeps like a baby. I am reminded of a line from Wordsworth: "Trailing clouds of glory do we come."

My father who loved to watch birds—nesting cardinals, blue jays, and flights of honking geese over Lake Michigan—is leaving now, returning to glorious clouds. The conversation with himself is finished. He rests above sky-tossed snows in the pure light he has always been seeking.

Still, we cannot leave him untouched. Each of us, in her own way, will try to offer some final comfort—if not to him, to ourselves. I find myself singing along with a Bach fugue in my off-key broken voice. I am stroking my father's bare shoulder, kissing his forehead, weeping. I tell him that in our next lives we will both be musicians—me, the singer, him, the piano player. I believe he senses my presence and is glad. Singing and kisses are the only gifts I have to offer.

Kathy is struck by the need to offer physical comfort, to witness the actual body as it dies. She pulls his sheets aside, examines the bedsores, the positions of his limbs. Our father's yellowish swollen hands are clenched. She gently pries open his fingers, rolls a soft washcloth into the palm of his hand to ease the pressure. She places a pillow under his bruised heel. Kathy looks down at her father's ruined body and does not flinch. Her tears wet his hands.

Carole is out in the hallway berating a nurse. We've just learned that contrary to our instructions, the nurse has been jamming a large plastic syringe down my father's throat. She fills it with milky porridge, then forces the food into him. She is from India. She does not understand how a man's children can allow him to starve when food is at hand. But Carole says, "No. You must not hurt him anymore." Carole comes to her father's death outraged. She will tolerate no invasion of his dignity. Her way to offer comfort is to protect his independence to the last breath. She crosses her arms, fierce with love.

A couple of miles away at Illinois Masonic my mother is sitting up now, able to walk, able to eat and drink, although food gives her no pleasure. Pleasure, I say, is not the point.

"How is your father?" she asks.

"He's at peace. There's no more pain."

"Good," she says. And she squeezes my hand. Then I am flying west above the clouds, chasing the sun. I feel no grief, only a letting go.

The hospice people had given us their handbook of dying. It is a thin, blue-covered pamphlet titled *The Ship Sails On.* Their wise and common language assures me that my father is ready for his last voyage, ready to sail away. His life was long and full, and he'd had enough of it. I am glad to have witnessed the struggle and the suffering, the turning inward and the release. And I hope that when my time comes, it will come slowly enough for me to also live fully my death.

Life, as the Buddhists have written in their Book of the Dead, can be understood as a long preparation for dying well. A spiral. A comet in galaxies. My father's last sculpture, which he named *The Reaper,* stands unfinished on his workbench. It is a shrouded figure with legs as thin and bowed as its maker's. Chisel marks are visible in pale naked wood, and death's blank face is buried within a caul that is delicate, like the membrane clinging to a fetus at birth. This is our father's legacy. A final statement created in pain and carved into wood. It is beautiful.

In the night, as I return to my sons and my lover, my father passes away. It is January 7, 1997, a week before his eighty-ninth birthday. I recall the old wise words he had written on a scrap of paper in his wallet. *The way is all.* And next to pictures of ourselves, another scrap, another poem. *We keep all we lose and love.*

THANKSGIVING

I look down at Chicago's gray apartment blocks and my eye is caught by lakeshore parks that sweeten the map of childhood like frosting. There is the green fan of Wrigley Field, where my father taught me to cheer losers; and there, the Lincoln Park Zoo, where I learned the caged habits of gorillas; and at the edge of the towering Loop are the lions of the Art Institute, where I discovered that life and art and nature are one thing.

From my thirty-second-floor perch in Mother's retirement home, it is impossible to see Warren Dunes State Park across Lake Michigan. But that's where we are going—where we've gone for over half a century—driving some eighty miles along the Indiana Tollway and U.S. 94, past the stinking refineries of Calumet City and the steel mills of Gary, to the hamlet of Sawyer on Michigan's cupped southwestern shore.

Mister Warren's dunes are drawing us toward an intimacy that cannot be found in the city. They sprawl along the crescent beach in mounds and hollows shaped by glaciers and scoured by wind and water. The tawny sand is 90 percent quartz, fine and sun warmed, sensuous as a ripe woman or a great-chested man. Behind the dunes rise hills of oak and maple, tulip tree, ash and pine, with clearings where we can lie down on layers of soft leaves, safely protected from the lightning storms that sweep inland across the waves any season of the year.

People come to this state park as they would come to a lover, seeking the solace and wildness they do not find in domestic situations. They may splash in July's waters with sunburned hordes or walk December's deserted beaches in snowy spindrift. There are red-leaved oaks and yellow cottonwoods in October, and an orange-fruited vine called bittersweet. And in spring there's a welter of wildflowers: jack-in-the-pulpit, Dutchman's-breeches, the white trillium with its delicate heart.

By June, small fish called alewives litter the beaches. They are putrid, flyblown reminders of mortality soon dried and returned to dust, but renewal is in the warm rain, in the blossoming chokecherry brush, and in green-lobed sassafras leaves. And a June walker may pick reddish purple blossoms from the papaw tree, a member of the custard apple family that spreads south into the Tropics.

I have never eaten its three- to five-inch fruit, which ripens in October and tastes like banana and pear. That's because the opossums, squirrels, foxes, and raccoons get there first. My father loved raccoons. He would feed them from his hand, pooh-poohing the notion they were infected with rabies.

When I was a child, I walked game trails through the woods, but never saw a deer. Now they have returned in such numbers that people may hunt them in the park's northern sections. Mother and I spotted two whitetails when we drove in. It was a metal-bright November day—surf roaring, the breeze scented with dried leaves and smoke. Gulls flocked on the sand, but the mallards and Canada geese had long since migrated south.

In the parking lot by the boarded-up, pagoda-shaped concession stand, a pigtailed girl who reminded me of me stamped her foot. "I will not leave until I climb Old Baldy!" she cried. Me too, I thought.

Eight middle-schoolers from Kyoto climbed ahead of me. Chattery as crows, they swung with their blond, blue-jeaned Michigan hosts from cottonwood snags on top of the 240-foot dune. Looking east, I saw a red-brown roof of trees that masked orchards, vineyards, and towns, creating the illusion of primeval America. To the west was the whitecapped lake. South lay our summer home on Tower Hill, but northward, toward the domed nuclear plant in Bridgman, the Great Warren Dunes Natural Area opened barren and unpeopled.

I had stepped onto the nearly vertical slope of a U-shaped blowout and was sliding in deep sand when I heard a loud rustle from below. I scanned a patch of cottonwoods skirted by pale, waving marram grass, but saw nothing. Then I heard the sound again and a black shape emerged into sunlight.

It was a huge, heavy bird. The bird unfolded his wide black wings. He spread his unmistakable half-moon tail. I saw a long neck and red wattles. I held my breath. The rooster sailed stately over the opposite dune and into the oaks. He was the first wild turkey I had ever seen in the dunes. Tears came to my eyes. "Turkey!" I shouted. I was shouting thanksgiving to no one, and to all creation.